NEONATAL MANUAL

Neonatal Manual

Edited by

CHIU Man-chun
MBBS(HK), FRCP(Edin), FRCP(Glasg)

and

FOK Tai-fai
MBBS(HK), FRCP(Edin)

The Chinese University Press

ISBN 962–201–544–1

THE CHINESE UNIVERSITY PRESS
The Chinese University of Hong Kong
SHATIN, N.T., HONG KONG

This title was published with a subsidy from
MEAD JOHNSON NUTRITIONALS.

Printed in Hong Kong by Magnum (Offset) Printing Co., Ltd.

Preface

Undoubtedly neonatology has evolved into a subspecialty of its own. The technological skills and monitoring equipments are highly sophisticated. Much of its work is of demanding and urgent nature, such that readily available data and references seem indispensable and a pocket management handbook is most helpful.

It is because of the above reason that we have produced the present Neonatal Manual. The materials that are required have become so great that it is no longer suitable for them to be contained in a paediatric manual. Thus, when revising the fifth edition of the Paediatric Manual, a separate manual in neonatology is published. It is accomplished with the conjoint effort of all those working in the field in Hong Kong. We hope that it will not only serve to help doctors and nurses managing patients in the neonatal wards, but also unifying practices in neonatology in our hospitals.

The Editors

Foreword

Dr Chiu Man-chun is the editor of the *Paediatric Manual* which has enjoyed a wide circulation and is now into its fifth edition, and Dr Fok Tai-fai is one of the most experienced and respected neonatologists in Hong Kong. This *Neonatal Manual*, a product of their expert collaborative efforts, is therefore assured of success and a long future. It does not really need a Foreword by anyone else but they have been gracious enough to invite me to contribute a few words, and I am delighted to accept.

Delighted, because I believe neonatology is coming of age in Hong Kong. Neonatology, as a subspecialty of paediatrics, holds a predominant position in a young doctor's postgraduate paediatric training programme as well as in a paediatrician's subsequent clinical practice, whether it be in the university, government or private sector. This book has been written to provide all doctors caring for the sick neonate, a ready reference to major neonatal disorders and their practical management. It is not merely a synopsis of what has been published in other neonatal textbooks. It reflects the state of the art in terms of how neonatology is being understood and practised in Hong Kong. It provides a general consensus and a common standard by which current neonatal protocols and procedures in Hong Kong should be used.

This book serves as a major milestone in the development of neonatology in Hong Kong. Twenty-one contributors collaborated in writing this *Neonatal Manual*. They belong to the new generation of specialists who possess the expertise and experience in the management of a variety of neonatal disorders. Together, five major hospitals are represented. Academic staff from both the University of Hong Kong and The Chinese University of Hong Kong made their contributions. Such collaboration not only ensures that this book reflects a proper balance of medical opinions but the editors deserve congratulations and encouragement for having built the bridges and narrowed the gulf separating major institutions. I do hope and pray that this will only be the beginning of a positive trend for closer ties between individuals and institutions dedicated to improving medical care to sick neonates and to improving their outcome, so that further cooperation can be fostered in a climate of mutual trust and respect.

Professor VICTOR YU
MD, MSc(Oxon), FRACP, FRCP(Lond), DCH
Director of Neonatal Intensive Care
Monash Medical Centre
Monash University, Melbourne
Australia
June 1991

Contents

Contributors

8. **HUEN Kwai-fun** (禤桂芬)   *18*
   MBBS(HK), MRCP(UK), DCH(Lond), DCH(Glasg), DCH(Ire)
   Senior Medical Officer
   Department of Paediatrics, Princess Margaret Hospital, Hong Kong

9. **KWONG Ngai-shan** (鄺毅山)   *9*
   MBBS(HK), MRCP(UK)
   Senior Medical Officer
   Department of Paediatrics
   Princess Margaret Hospital, Hong Kong

10. **LAM Kwun-lai, Paul** (林觀禮)   *10, 27*
    MBBS(HK), MRCP(UK), DCH(Lond)
    Senior Medical Officer
    Department of Paediatrics
    University of Hong Kong, Queen Mary Hospital, Hong Kong

11. **LEE Hon-kwan** (李漢鈞)   *6*
    MBBS(HK), MRCP(UK), DCH(Lond)
    Senior Medical Officer
    Department of Paediatrics, Princess Margaret Hospital, Hong Kong

12. **LEE Ngar-yee, Natalie** (李雅兒)   *1*
    MBBS(HK), MRCP(UK), DCH(Lond)
    Former Lecturer
    Department of Paediatrics
    The Chinese University of Hong Kong
    Prince of Wales Hospital, Hong Kong

13. **LEE Wai-hong** (李偉航)   *9, 25*
    MBBS(HK), MRCP(UK), DCH(Lond) , DCH(Glasg)
    Consultant Paediatrician
    Department of Paediatrics, Queen Elizabeth Hospital, Hong Kong

14. **LEUNG Chi-wai** (梁志偉)   *22*
    MBBS(HK), MRCP(UK), DCH(Glasg), DCH(Ire)
    Senior Medical Officer
    Department of Paediatrics, Princess Margaret Hospital, Hong Kong

15. **LEUNG Nin-ming** (梁念明)   *5, 11*
    MBBS(HK), MRCP(UK), DCH(Lond)
    Former Consultant Paediatrician
    Paediatric Unit, Tuen Mun Hospital, Hong Kong

16. **LEUNG Piu-ngor, Ellen** (梁佩娥)   *19*
    MBBS(HK), MRCP(UK), DCH(Lond)
    Former Senior Medical Officer
    Department of Paediatrics, Queen Elizabeth Hospital, Hong Kong

1

Infant Feeding, Fluid Requirement, and Parenteral Nutrition

Natalie NY Lee

1. Calories requirement

1.1 100–120 kcal/kg/day

2. Protein requirement (20%)

2.1 **1/12 of age:** 2.5 gm/kg/day
6/12 of age: 2.2 gm/kg/day
2.2 **Essential amino acids:** histidine, isoleucine, lysine, methionine, phenylalanine, threonine, tryptophan, valine, cystine (preterm).

3. Carbohydrate requirement (50%)

3.1 10–15 gm/kg/day
3.2 Ketosis if carbohydrate intake falls < 15% of total caloric intake.

4. Fat requirement (30%)

4.1 3–4 gm/kg per day
4.2 **Essential fatty acids:** linoleic acid, linolenic acid.

5. Suggested daily intake of macromolecules

5.1 **Phosphorus**
❏ 1 mmol/kg/day

5.2   Magnesium
   ❏ 0.3–0.5 mEq/kg/day
5.3   Iron supplement
   ❏ term   1 mg/kg/day   starts at 4/12 of age
   ❏ preterm   2 mg/kg/day   starts at 6–8 weeks.

6.   Vitamins requirement

Recommended daily intake

Nutrient	Term newborn	Preterm/Low birth weight
D	400 IU	800–1000 IU
A	420 mg (≈ 1300 IU)	500 IU
K	12 µg	1–1.5 mg at birth, repeat at 7–10 days interval if necessary
E	3 mg	5–50 IU
C	35 mg	50 mg

7.   Enteral feeding

7.1   Full term infants
   ❏ Feeding can be started 4–6 hours after birth.
   ❏ Full term healthy infants can be fed at interval of every 3–4 hours. By 6 months of age, the number of feeds can be reduced to 4 times daily.
7.2   Preterm infants
   ❏ For neonates < 1500 gm
     • Tube feeding initially because of inability to co-ordinate sucking and swallowing.
     • Start with 1/4 or 1/2 strength milk, increase either the strength or volume, one at a time.
     • Suggested feeding regime:

	Initial amount	Frequency	Increment
< 1000 gm	1–2 ml	1–2 hours	1 ml initially then 2 ml (max. 10–15 ml/feed)
1000–1500 gm	2–4 ml	2 hours	2–4 ml, depend on tolerance (max. 15–22 ml/feed)
1500–2000 gm	3–5 ml	2–3 hours	2–5 ml (max. 30–40 ml/feed)

❏ Note:
  • Weight gain may not be achieved initially until infants have been on full feeding.
  • Stomach contents must be aspirated every 4 hourly to make sure feeding are well tolerated.
  • Infants' stomach empty better if they lie on their right side or are lying prone.

8. Total parenteral nutrition

8.1 Introduction

❏ Babies with fulminating sepsis (such as NEC, septicaemia) should be adequately stabilized with antibiotic therapy before starting TPN.

❏ All solutions should be prepared under aseptic technique, preferably by the pharmacists under lamina flow unit. Infusion sets should be changed daily.

❏ Intake of large amount of fat emulsion is contra-indicated in neonate with
  • severe oxygenation defect
  • a plasma bilirubin > 12 mg% (200 µmol/L)
  • thrombocytopenia.

For such cases, give a small amount of lipid emulsion (0.5–1.0 g/kg/day) to prevent essential fatty acids deficiency which may develop within a few days of total lipid deprivation.

❏ Theoretically, carnitine is required for optimal oxidation of fatty acids.

❏ 20% lipid emulsions may be advantageous over 10% preparation because of its
  • smaller volume
  • lower phospholipid content leading to lower cholesterol level.

❏ Route of administration may be a peripheral or central venous line.
The central route should be used if the concentration of dextrose exceeds 10%.

❏ Mix dextrose solution with amino acid solution adding electrolytes, trace elements and soluble vitamins in one bottle and lipid emulsions with vitalipids in a separate bottle. The two mixtures should be administered simultaneously via a Y connector.

Table 1. Suggested TPN Scheme for Newborn & Infants (per kg body weight per day)

(Based mainly on Easton LB et al.: *Parenteral nutrition in the newborn: A practical guide.* Pediatr Clin N Amer 1982; 29:1171–90.)

Constituent	Starting	Daily increment	Full (maximum)	Side-effects
Dextrose 10%	Fluid requirement of the baby		See footnote*	Hyperglycaemia, glycosuria → osmotic diuresis & dehydration may need to decrease strength (7.5%, 5% etc.) when hyperglycaemia occurs.
Vamin Glucose (pH 5.2)	0.5 g A.A. (7 ml = 0.42 g protein, 0.66 g N)	0.5 g A.A. (7 ml)	2.5 g A.A. (35 ml)	Thrombophlebitis — better to administer simultaneously with intralipid through a joint catheter. Azotemia, acidemia, hyperammonaemia (rare with A.A. concentrates). Cholestatic jaundice.
Lipid 20% Intralipid (pH 7.5)	0.4–0.5 gm (2–3 ml)	0.5 gm (2.5 ml)	2.5–3 gm** (12.5–15 ml) (may need smaller amount in VLBW) (rate of infusion must not > 0.25 g/kg/hr)	May alter glucose metabolism and cause hyperglycaemia Early: fever, hypercoagulability thrombocytopenia (rare). Delayed: thrombocytopenia, leukopenia, hepatospleno-megaly, transient increase in transaminases and cholestatic jaundice. Do not mix with other drugs, nutrients or electrolytes.

Table 1. Suggested TPN Scheme for Newborn & Infants (per kg body weight per day) (continued)

Constituent	Starting	Daily increment	Full (maximum)	Side-effects
Ped-E1			4 ml	
Soluvit (add to Dextrose)			0.5 ml	
Vitalipid Infant (add to Lipid) (pH 8)			1 ml (Total amount must not exceed 4 ml)	
KH_2PO_4 + K_2HPO_4,* (1 ml = 2.5 mmol K + 1.45 mmol P) (add to Dextrose)		{Total amount phosphorus required (Normally 1 mmol/kg/d) − [vol. of Intralipid (ml) × 0.015 + vol. of Ped-E1 (ml) × 0.075]} mmol (Usually not given during 1st week of TPN when abnormalities in Ca homeostasis predominates.)		Abnormal deposition of Ca occurs if Ca-P product in serum > 70 (both in mg/dl).
Na		{Total amount Na required (Normally 2–3 mmol/kg/d) − [Vol. of Vamin (ml) × 0.05]} mmol		
K		{Total amount K required (Normally 2–3 mmol/kg/d) − [Vol. of Vamin (ml) × 0.02 + Vol. of K-P mixture (ml) × 2.5]} mmol		
Ca		{Total amount Ca required (Normally 1 mmol/kg/d) − [Vol. of Vamin (ml) × 0.0025 + Vol. of Ped-E1 (ml) × 0.15]} mmol		Bradycardia, cardiac arrhythmia + CNS depression if hypercalcaemic

* To allow optimum utilization of amino acids, when 2.5 gm A.A./kg/d is being given, a minimum of 60 cal/kg/d of glucose must be provided. During full TPN when 35 ml/kg/d of Vamin being given, the amount glucose inside Vamin = 3.5 g = 14 kcal. Hence the D_{10} solution must give (60–14) kcal = 46 kcal, then the amount of D_{10} required must > 115 ml/kg/d.

** Manufacturer's recommendation: can go up to 4 gm/kg/d.

*** When the normal requirement of Ca & P is added to a full TPN preparation, precipitation normally does not occur. When the additional amount of either is required or when the volume of the TPN preparation is decreased, may need to give Ca & P separately, each for 12 hours.

Table 2. Suggested TPN Scheme if Other Amino Acid Solutions (Trophamine/Aminoplasmal^R Paed) Are Used

Solutions	Starting	Daily increment	Full (maximum)	Total TPN fluid	kcal/kg/day
Dextrose 12.5%*	Fluid requirement of the baby		120 ml		
Trophamine (pH 7.5)	0.5 gm A.A. (8 ml = 0.46 gm protein)	0.5 gm A.A. (8 ml)	2.5 gm A.A. (40 ml)	175 ml	99.6
OR					
Amino plasmal Paed	0.5 gm A.A. (10 ml)	0.5 gm A.A. (10 ml)	2.5 gm A.A. (50 ml)	185 ml	100
20% Lipids	0.4–0.5 gm (2–3 ml)	0.5 gm (2.5 ml)	2.5–3 gm (12.5 ml–15 ml)		

* Since Trophamine and Aminoplasmal^R Paed do not contain glucose, so 12.5% Dextrose is recommended to allow the optimum utilization of amino acids, i.e., when 2.5 g A.A. is being given, a minimum of 60 kcal/kg/day of glucose must be provided.

Table 3. Constituents of Various Solutions

Amino acid solutions

	Vamin^R with glucose	Trophamine	Aminoplasmal^R Paed
100 ml contain:			
Total amino acids	~ 70 g	60 g (+ taurine)	50 g
Total nitrogen	9.4 g	9.3 g	7.4 g
Glucose	100 g		
Energy	650 kcal	240 kcal	200 kcal
Sodium mmol (m Eq)	50	5	45
Potassium mmol (m Eq)	20		25
Calcium mmol (m Eq)	2.5 (5)	< 3	
Magnesium mmol (m Eq)	1.5 (3)		2.5 (5)
Chloride mmol (m Eq)	~ 55		15
Osmolality mosm/kg	~ 1350	525	?

Vitamins

Soluvit^R		*Vitalipid*^{R} *Infant*	
1 vial (10 ml) contains:		**1 ml contains:**	
Vit C	30 mg	Vit A	100 mg (333 IU)
B$_1$	1.2 mg	D$_2$	2.5 µg (100 IU)
B$_6$	2.0 mg	K$_1$	50 µg
Nicotinamide	100 mg		
Pantothenic acid	10 mg		
Biotin	0.3 mg		
Folic acid	0.2 mg		
Vitamin B$_{12}$	2.0 mg		

Ped-el^R

1 ml contains:

Ca	0.15 mmol	F	0.75 µmol
Mg	25 µmol	I	0.01 µmol
Fe	0.5 µmol	P	75 µmol
Mn	0.25 µmol	Cl	0.35 mmol
Zn	0.15 µmol	Sorbitol	0.3 g
Cu	0.075 µmol		

8.2   Monitoring of IV feeding
❑ Daily
  • Body weight
  • Urine sugar
    Dextrostix q12h
    Blood sugar prn
  • Lipaemia:
    – Quantitative analysis of plasma lipid profile.
    – Centrifuge capillary blood to note whether supernatant
      serum is turbid. If lipaemia, stop lipid and restart with half
      of the previous amount after the plasma has become clear.
      This is a crude method; absence of visually detectable
      lipaemia does not rule out hyperlipidaemia.
  • Blood gases
    Unexplained metabolic acidosis: check for lipaemia, lactic
    acidosis and infection.
  • Urea and electrolytes
    Abnormal electrolyte levels are abnormal: check for lipaemia.
    Increased urea:
    – Check if intake of fat/carbohydrate is adequate.
    – Check for infection.
❑ Weekly
  • LFT, triglycerides, cholesterol
  • Haemoglobin, WBC, platelet
  • Blood culture if indicated

References

1.   Avery GB. *Neonatology*. 3rd ed. Philadelphia, WB Saunders, 1987.
2.   Roberton NRC (ed). *Textbook of Neonatology*. 1st ed. Great Britain,
     Churchill Livingstone, 1986.

2

Fluid and Electrolyte Disturbances

CHIU Man-chun

1. Fluid therapy

1.1 Initial requirement[1,3]

$D_{10}W$ iv at 60 ml/kg/24 hr on Day 1 and increased to 150 ml/kg/24 hr on Day 5

1.2 Insensible water loss (IWL)

Measurements (ml/kg/hr)

Weight of infant (kg)	Incubator	Radiant warmer bed
0.6–1.0	1.5–3.5	2.4–5.2
1.0–1.5	1.5–2.3	1.5–2.7
1.5–2.0	0.7–1.0	0.5–1.5
3.0	0.5	1.0

Factors Affecting IWL

Increases IWL (%)	Decreases IWL (%)
Severe prematurity (100–300)	Humidification in incubator (50–100)
Open warmer bed (50–100)	Plastic heat shield incubator (30–50)
Forced convection (30–50)	Plastic blanket under radiant warmer (30–50)
Phototherapy (30–50)	Tracheal intubation with humidification (20–30)
Hyperthermia (30–50)	
Tachypnea (20–30)	

1.3 In RDS[2]

❑ Initial stabilization
  • For hypoperfusion or hypotension: blood or saline 10–20 ml/kg over 30–60 min.
  • For severe acidosis pH ≤ 7.2, or serum HCO_3 ≤ 15 mmol/l, $NaHCO_3$ 1–2 mmol/kg iv slowly in 10–15 min.
  • For hypoglycaemia: increase D10 infusion rate two- to threefold until blood glucose returns to normal.
❑ Maintenance with restriction
  Relative fluid restriction with no NaCl for several days until RDS improving
❑ Liberal fluid
  When diuresis sets in with improvement of pulmonary function, increase fluid to ≥ 120 ml/kg/day.

1.4 In perinatal asphyxia

Fluid restriction of 40–70 ml/kg/day and avoiding rapid bolus infusion may help alleviating cerebral oedema.

2. Sodium

2.1 Supplement/requirement [4]

❑ None in first 3 days
❑ BW > 1,250 gm, 2–3 mmol/kg/day
❑ BW ≤ 1,250 gm, 4 or more mmol/kg/day
❑ Rapidly growing premature, 4–5 mmol/kg/day

Note: For VLBW infants, more sodium might be required especially after 1st week of life because of excessive urinary sodium loss.

2.2 Inadvertent gain/loss

❑ Gain: bicarbonate, heparin, antibiotics, saline used for flushing catheters
❑ Loss: blood sampling, nasogastric aspiration, frequent LPs or cerebral ventricular taps

2.3 Hyponatraemia (Na < 130 mmol/l)

❑ Common causes
  • Inadequate intake
  • Excessive GI loss, e.g. diarrhoea, NG aspiration
  • Inappropriate ADH
  • Diuretics
  • Adrenal insufficiency
  • Congenital adrenal hyperplasia
  • Increased natriuresis in VLBW due to immaturity of angiotensin-renin-aldosterone system[5]
  • Acute renal failure

❑ Replacement

[Na] required (in mmol) = (desired [Na] − present [Na]) × 0.6 × wt (kg)

2.4 Hypernatraemia (Na > 150 mmol/l)

❑ Common causes

- Fluid restriction with excessive water loss
- Excessive Na intake parenterally
- Faulty milk dilution
- Excessive $NaHCO_3$ for correction of acidosis
- Diarrhoea with inappropriate replacement
- Diabetes insipidus

❑ Treatment

According to cause; however should correct gradually at a rate of reduction of [Na] 10–15 mmol/l daily. In cases of severe dehydration or shock, after plasma volume repletion, ½ : ½ solution (fluid with 75 mmol/l Na) should be given until good urine output is established. Then more hypotonic solution can be used.

3. Potassium

3.1 Hypokalaemia (K < 2.5 mmol/l)

❑ Common causes

- GI loss
- Inadequate supplement
- Drugs; diuretics, dopamine
- Respiratory alkalosis: e.g. Rx of persistent pulmonary hypertension
- Bartter's syndrome
- Renal tubular acidosis
- Adrenal causes: e.g. excessive mineralocorticoids

❑ Treatment

KCl 2–4 mmol/kg/day; avoid bolus infusion

3.2 Hyperkalaemia (K > 6 mmol/l)

❑ Common causes

- Falsely high due to haemolysis in collection, or in vitro haemolysis due to leukocytosis, thrombocytosis
- Haemorrhage: intracerebral, intrapulmonary, severe bruises due to birth trauma
- Excessive infusion of KCl
- Renal failure
- Adrenal causes: CAH, mineralocorticoid deficiency

❑ ECG changes

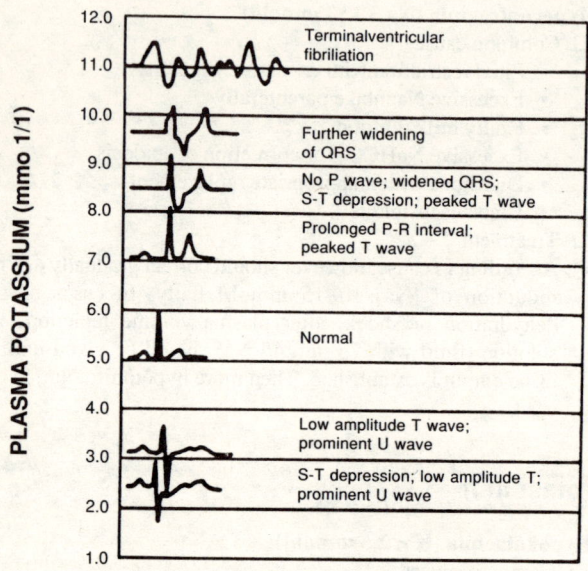

The relationship between plasma potassium concentration and electrocardiographic changes.

❑ Treatment
- Expectant
  - Check iv fluid, and recheck serum K
  - ECG monitoring
  - cat-ion (Na or Ca) exchange resin (Kayexelate) 0.5–1.0 gm/kg every 6–12 hrs
- With ECG changes
  - NaHCO$_3$ 1 mmol/kg iv over 15 min
  - Ca gluconate (10%) 0.5–1.0 ml/kg iv over 15 min
  - Glucose (increase 50–100% D10 infusion rate) plus insulin 0.1–0.2 units/kg iv or sc every 4–6 hrs
- Dialysis/haemofiltration
  - When above measures fail, dialytic treatment may be required
- For arrhythmias
  - Cardioversion or lidocaine iv 1 mg/kg

4. Calcium

4.1 Hypocalcaemia (Ca < 1.85 mmol/l or ionized Ca < 0.9 mmol for term baby)

❏ Common causes
- Early/Transient
  - Prematurity
  - Asphyxia
  - Infant of diabetic mothers
  - Infant of hyperparathyroid mothers
  - Post exchange transfusion
- Late/Prolonged
  - DiGeorge syndrome
  - Transient congenital idiopathic hypoparathyroid
  - Hypoparathyroidism/pseudohypoparathyroidism
  - Magnesium dependent hypocalcaemia
  - Unmodified cow's milk
  - Renal failure
  - Rickets of prematurity

❏ Manifestations
- Irritability, apnoea, seizures
- ECG changes: prolonged QTc

❏ Treatment
- **Symptomatic**
  - Acute: 10% Ca gluconate 1–≤ 2 ml/kg iv slowly at ≤ 1 ml/min; may be repeated 3–4 times a day.
  - Maintenance: 10% Ca gluconate 5–10 ml orally or iv in 24 hrs; adjusted accordingly to serum level.
- **Asymptomatic**
  - Oral Ca supplementation
  - Vit D 400 IU/day
- **Prolonged/2° to hypoparathyroidism**
  - Low P milk (Similac PM 60/40)
  - Oral Ca supplementation
  - P binders: Ca carbonate
  - Vit D — Calciferal 1,000–3,000 IU/kg/d (or 0.025–0.075 mg/kg/d)
  - 1,25 dihydroxyvit D3: 20–50 ng/kg/day

4.2 Hypercalcaemia (Ca > 2.75 mmol/l) (or ionized Ca > 1.25 mmol/l)

❏ Common causes
- Iatrogenic: overcorrection of hypocalcaemia with excessive iv Ca supplementation
- Phosphate depletion
- Primary hyperparathyroidism
- Maternal hypoparathyroidism

- Subcutaneous fat necrosis syndrome
- Williams syndrome

❏ Manifestations
Poor feeding, vomiting, hypotonia, lethargy, polyuria, dehydration, hypertension, seizures.

❏ Treatment[8]

Treatment	Dose	Mode of action
SYMPTOMATIC		
Normal saline	1.5–3 × maintenance for age	Augmented urinary Ca excretion
Frusemide	1 mg/kg/dose q6h, if necessary	Augmented urinary Ca excretion
Calcitonin	4 MRC units/kg/dose q12h s.c.	Inhibition of bone resorption
ASYMPTOMATIC		
Discontinue vitamin D supplementation		
Restrict dietary Ca		
Avoid sunlight		
Consider parathyroidectomy (subtotal) in cases of primary neonatal hyperparathyroidism		

5. Magnesium

5.1 Hypomagnesaemia (Mg < 0.6 mmol/l)
It should be suspected whenever hypocalcaemia is symptomatic or persistent despite adequate calcium supplementation. It can be treated with $MgSO_4$ 100 mg/kg in every 12 hr for 3–4 doses, and if persistent, prolonged oral therapy is necessary.

5.2 Hypermagnesaemia (Mg > 1.2 mmol/l)
It should be suspected when mother given magnesium iv for toxaemia, and baby manifests apnoea, bradycardia, hypotension, hypotonia and areflexia. Treatment is mainly supportive, though Ca gluconate infusion, exchange transfusion, or dialysis may help.

6. Phosphorus

6.1 Hypophosphataemia (< 1.6 mmol/l [term])
Common causes include prolonged parenteral alimentation with phosphate depletion, Fanconi syndrome and hyperparathyroidism.

6.2 Hyperphosphataemia (> 2.2 mmol/l [term])

Common causes include renal failure, massive phosphate infusion, and hypoparathyroidism/pseudohypoparathyroidism.

References

1. Avner ED, Hazinski TA, Dabbagh S et al. *Fluid and Electrolyte Abnormalities in the Stressed Neonates.* In Ichikawa I (ed): *Pediatric Textbook of Fluids and Electrolytes.* Baltimore, Williams & Wilkins, 1990; pp. 402–13.

2. Costarino A, Baumgart S. *Modern fluid and electrolyte management of the critically ill premature infant.* Pediatr Clin North Am 1986; 33(1):160.

3. Roberton NRC. *A Manual of Neonatal Intensive Care.* 2nd ed. London, Arnold ELBS, 1986; pp. 34–35.

4. Berry PL, Belsha CW. *Hyponatremia.* Pediatr Clin North Am 1990; 37(2):360.

5. Sulyok E, Varga, Goyory E et al. *Postnatal development of sodium handling in premature infants.* Pediatr 1982; 95:787.

6. Satlin LM, Schwartz GJ. In Ichikawa I (ed): *Pediatric Textbook of Fluids and Electrolytes.* Baltimore, Williams & Wilkins, 1990; pp. 218–36.

7. Guignard JP, John EG. *Renal function in the tiny, premature infant.* Clin Perinatol 1986; 13(2):377–402.

8. Carpenter TO, Key LL. *Disorders of the Metabolism of Calcium, Phosphorus, and Other Divalent Ions.* In Ichikawa I (ed): *Pediatric Textbook of Fluids and Electrolytes.* Baltimore, Williams & Wilkins, 1990; pp. 237–68.

9. Anast CS. *Disorders of Mineral and Bone Metabolism.* In Avery ME, Taeusch HW (eds): *Schaffer's Diseases of the Newborn.* 5th ed. Philadelphia, WB Saunders, 1984; pp. 469–70.

3
Hypothermia

Robert KN YUEN

1. Introduction

1.1 **Definition:** Core body temperature < 35°C.
1.2 Hypothermia markedly increases mortality in sick and very low birthweight newborns.
1.3 Cold stress usually occurs in labour ward, operation theatre and during transport.

2. Deleterious effects of hypothermia in LBW babies

Inhibit surfactant production
Pulmonary hypertension and subsequent hypoxia
Metabolic acidosis
Hypoglycaemia
Increased oxygen consumption
Clotting defect and pulmonary haemorrhage

3. Practical guide for prevention

3.1 Keep labour ward and neonatal wards warm (26–28°C) and draught free.
3.2 Prewarm transport incubator and overhead radiant warmer before use.
3.3 Newborn should be dried thoroughly, including the head, immediately after birth and well wrapped up with warm, dry blankets.
3.4 Warm and humidify medical gas delivered to babies.
3.5 Place babies under radiant warmer for resuscitation and in prewarmed incubator for transport.

3.6 Do not bathe babies after delivery until temperature has been stabilized.

3.7 Adjust incubator temperature to provide the neutral thermal environment.

3.8 Use double-glazed incubators and cover the baby with insulating fabric (e.g. silver swaddlers, clingfilm) and radiant heat shield.

3.9 Insulate babies being nursed under radiant warmer on an open bassinet with transparent plastic blanket but not perspex heat shield.

3.10 Cover baby's head with cap and place heating pad set at 37°C under the baby.

3.11 For babies requiring ventilator care, humidify inspired gas and ensure the temperature of inspired gas is 35–37°C.

3.12 Set abdominal skin temperature at 36.5°C for servo-control mode; beware of overheating which occurs when sensor is loosened.

Table 1. Neutral Thermal Environmental Temperatures

Approximate neutral thermal environmental temperatures (in degree celsius)[a]

Age	1200 gm or less	1201–1500 gm	1501–2500 gm	More than 2500 gm or more than 36 weeks
0–24 hours	34–35.4	33.3–34.4	31.8–33.8	31–33.8
24–48 hours	34–35	33–34.2	31.4–33.6	30.5–33.5
48–72 hours	34–35	33–34	31.2–33.4	30.1–33.2
72–96 hours	34–35	33–34	31.1–33.2	29.8–32.8
4–14 days	DNE[b]	32.6–34	31–33.2	29–32.6
2–3 weeks	DNE	32.2–34	30.5–33	DNE
3–4 weeks	DNE	31.6–33.6	30–32.7	DNE
4–5 weeks	DNE	31.2–33	29.5–32.2	DNE
5–6 weeks	DNE	30.6–32.3	29–31.8	DNE

[a] In general, the smaller the infant, the higher the temperature required.

[b] Data not established.

DNE = Data from Scopes J. Ahmed I, *Range of initial temperatures in sick and premature newborn babies*. Arch Dis Child 1966; 41:417.

4

Complications of Prematurity and Dysmaturity

Grace LH CHAN

1. Definitions[1-4]

1.1 **Low birth weight (LBW):** birth weight < 2500 g.
Very low birth weight (VLBW): birth weight < 1500 g.
Extreme low birth weight (ELBW): birth weight < 1000 g.

1.2 **Prematurity:** gestation < 37 weeks

1.3 **Small for gestational age (SGA)**
- ❏ Intrauterine growth retardation (IUGR)
- ❏ Less than 10th centile for gestational age *or* more than 2SD below the mean for gestational age
  - Symmetric IUGR:
    OFHC, Length, Birth weight all < 10%
  - Asymmetric IUGR:
    OFHC, Length > Birth weight but all < 10%

1.4 **Categories of LBW**
- ❏ 2/3 born early, i.e. prematurity
- ❏ 1/3 born small, i.e. SGA
- ❏ Minority, prematurity + SGA

2. Complications of prematurity

2.1 **Early complications**
- ❏ Instability of body temperature/Hypothermia
  Causes:
  - Relatively large surface area
  - Lack of subcutaneous fat
  - Lack of muscular activity

- Immature heat regulatory centre
❑ Respiratory
  - Recurrent apnoea/bradycardia
    - Confined to LBW < 1.5 kg or < 32 weeks gestation
    - 28–30 weeks gestation: 100% incidence
    - Onset may be delayed to Day 2–6
  - Respiratory distress syndrome (R.D.S.)[6]
    10–15% incidence (28–30 weeks: 70% affected)
  - Wilson-Mikity Disease[5,7]
    - Typically seen in VLBW infants
    - Insidious onset of respiratory distress at 1st–5th weeks of age
    - CXR: Honeycomb appearance
❑ Metabolic
  - Hypoglycaemia
    - 15% incidence within first 72 hours
  - Hypocalcaemia
  - Hyperbilirubinaemia[5]
❑ Neurological
  - Periventricular haemorrhage
    - 30–50% incidence in VLBW babies
    - Majority occurs in first 72 hours after birth
    - 50% have no clinical manifestations
  - Hypoxic Ischaemic Encephalopathy[8]
    - The commonest cause of seizures during intensive care of VLBW infants
❑ Persistent ductus arteriosus
  - 10–15% in LBW infants 1.5 kg–2 kg
  - > 80% in ELBW infants
  - May present as cardiac failure, apnoeic episodes or need for more vigorous mechanical ventilation
❑ Oedema/Metabolic acidosis[9]
  - Related to increased capillary permeability, hypoproteinaemia and poor renal function
❑ Susceptibility to infection/Septicaemia[3]
  - Evidence of maternal chorioamnionitis in a large proportion of premature deliveries
  - Transfer of maternal immunoglobulins becomes significant only after 28 weeks
  - Infection may develop in an atypical fashion at an alarming speed
    - Jaundice unexpectedly high/prolonged
    - Sudden change/instability in body temperature
    - Refuse feeding or slow sucking
    - Poor weight gain

       – Dullness, listlessness or irritability
- ❏ Gastrointestinal
  - • Functional ileus:[8]
    Rhythmic peristaltic activity may not appear until sucking and swallowing has been achieved at 34 weeks gestation
  - • Necrotising enterocolitis (NEC)
- ❏ Iatrogenic[8] — Common in neonatal intensive care unit. Examples include:
  - • Drug toxicity
    e.g. Tolazoline: gastric ulceration, haemorrhage
    e.g. Calcium gluconate extravasation: Dystrophic calcification
  - • Mechanical ventilation
    - – Trachael Intubation:
      Ulcer of nasal skin, palatal groove, central cleft, enamel defects, laryngeal/trachael ulceration and stenosis.
      Cutaneous excoriation and facial scarring caused by adhesive strapping.
      (Occurs in 1.5% of infants who have been intubated for > 4 weeks.)
    - – Barotrauma:
      Interstitial emphysema
      Pneumothorax (20% with IPPV)
- ❏ Vascular catheter
  - • Peripheral artery cannulation:
    - – Thrombosis, gangrene, scarring and pigmentation of extremities
  - • Umbilical arterial catheterization:
    - – Gangrene of buttock and lower limbs
  - • Umbilical venous catheterization:
    - – Pulmonary/Hepatic/IVC thromboembolism
    - – NEC
    - – Portal hypertension
- ❏ Total parenteral nutrition
  10–30% develops cholestasis ± portal fibrosis (? related to Amino Acid infusion)

2.2 Late complications
- ❏ Anaemia
  - • Early anaemia of prematurity
    - – Normochromic
    - – Decreased rate of erythropoiesis and relatively rapid rate of growth
  - • Late Anaemia of prematurity
    - – Iron deficiency: Hypochromic, 3–4 months old
    - – Folate deficiency: Macrocytic, 2–3 months old
    - – Vitamin E deficiency: Haemolytic, 6–10 weeks old

- Copper deficiency: Normochromic Normocytic, Neutropenia
❑ Metabolic bone disease of prematurity
  • 60% incidence in ELBW infants
  • X-ray grading[10]
    - grade 0: Normal bones
    - grade I: Rarefaction
    - grade II: Rachitic changes at bone ends
    - grade III: Rachitic changes + fractures
❑ Oxygen Toxicity
  • Retinopathy of prematurity[2,6]
    - Incidence:
      VLBW 10–35%
      ELBW 20–80%
    - Fundoscopic examination: 4 stages
      (i) Demarcation line
      (ii) Ridge formation
      (iii) Extra retinal fibrovascular proliferation
      (iv) Retinal detachment
      ("Plus" disease: Vessel dilatation + tortuosity)
    - Features: mild cases, myopia; severe cases, glaucoma usually bilateral
    - Screening: optimal timing: 4–6 weeks old
  • Bronchopulmonary dysplasia (B.P.D.)[6]
    - Incidence: 15–35% of VLBW infants, 5% of all NICU admissions requiring IPPV
    - Definition[8]
      (i) IPPV during the first week, lasting ≥ 3 days
      (ii) Chronic respiratory distress persisting > 28 days
      (iii) Oxygen therapy > 28 days
      (iv) X-ray changes: strands of increased density alternating with areas of increased lucency
    - Mortality: 30–40% during first year
    - Prognosis:
      (i) During first year: repeated episodes of lower respiratory tract infections with wheezing or evidence of reactive airway disease
      (ii) Recovery may take as long as 2 years
      (iii) Abnormalities in pulmonary function often persist for years even though asymptomatic
❑ Post-haemorrhagic Hydrocephalus/Porencephaly[13]
  • Occurs in > 20% patients surviving PVH (severe PVH 65–100%)
❑ Periventricular leucomalacia[13]
  • Apparent on cerebral ultrasound by 6–10 weeks of age

- Shown up initially as periventricular echodense shadows and later as periventricular cystic areas
- 2–8% incidence in VLBW infants, secondary to ischaemia
❏ Cerebral palsy
  - Spastic diplegia commonest
  - Overall incidence 3–6%
❏ Mental retardation/Developmental delay
  - Overall incidence 10–12%
❏ Deafness
  - Incidence 1–3%, usually sensorineural
❏ Impaired learning/Hyperactive behaviour
❏ Impaired mother — infant bonding
❏ Increased incidence of sudden infant death

3. Complications of dysmature/SGA infants

3.1 Intrapartum
❏ Intrauterine death[5] due to acute/chronic hypoxia
❏ Meconium aspiration

3.2 Postpartum
❏ Birth asphyxia[1,2]
❏ Metabolic
  - Hypoglycaemia: occurs in 15–65% especially in the first 12 hours
  - Hypocalcaemia — Idiopathic
❏ Temperature instability
❏ Polycythaemia/Hyperviscosity syndrome
  - Venous Hct > 65%
  - Complications
    - Jaundice
    - Renal venous thrombosis
    - Cardiac failure
    - NEC
    - Tissue infarction particularly in the brain and gut
❏ Respiratory
  - Persistent pulmonary hypertension[4]
    (Persistent fetal circulation)
    Secondary to chronic intrauterine hypoxia
  - Pulmonary haemorrhage
    may be massive
❏ Neurological/behavioural/feeding problems
  - altered reflex activity, poor muscle tone, jitteriness
  - less active/responsive to social stimuli
  - unpredictable feeding/sleeping patterns

- ❏ Susceptibility to infection
  Impaired cell mediated and humoral immunity
- ❏ Congenital infections — may be the cause of the intrauterine growth failure
- ❏ Congenital malformations
  10–20 times higher incidence[11]
- ❏ Development and growth[4]
  - Neurodevelopmental outcome:
    Symmetric IUGR — usually poor
    Asymmetric IUGR — usually good
  - Growth: Postnatal weight loss 0 to < 5%, followed by rapid gain. Catch up growth within first 6 months to 1 year or remain small

References

1. Roberton NRC. *Textbook of Neonatology.* 1st ed. Churchill Livingstone, 1986.
2. Fanaroff A, Martin J. *Neonatal — Perinatal Medicine, Diseases of the Fetus and Infant.* 4th ed. Mosby, 1987.
3. Hoekelman A. *Primary Pediatric Care.* 1987; pp. 549–52.
4. Gomella, Cunningham. *Neonatology, Basic Management, On-call Problems, Diseases, Drugs.* In *A Large Clinical Manual.* 1988; p. 121, p. 281.
5. Roberton NRC. *A Manual of Neonatal `Intensive Care.* 2nd ed. London, Arnold, 1986.
6. Sweeney K. *The High Risk Neonate, Developmental Therapy Prospectives.* The Haworth Press, 1986.
7. Cloherty P, Starch R. *Manual of Neonatal Care.* 1st ed. 1980.
8. Harvey D, WI Cooke, Levitt A. *The Baby Under 1,000 g.* 1st ed. Batterworth, 1989.
9. Vulliamy DG. *The Newborn Child.* 3rd ed. Churchill Livingstone, 1973.
10. Koo WWK, Gupta JM, Nayanar VV et al. *Skeletal changes in preterm infants.* Arch Dis Child. 1982; 57:447–52.
11. Klau H. *Care of the High Risk Neonate.* 1979; pp. 82–84.
12. Hodson WA and Truog WE. *Critical Care of the Newborn.* 2nd ed. Saunders, 1989; pp. 24–26.
13. Bernhaum JC, Friedman S, Hoffman-Williamson M et al. *Preterm infant care after hospital discharge.* Pediatrics in Review 1989; 7(7):195–206.

5

Resuscitation of Newborn

LEUNG Nin-ming

1. General guidelines

1.1 Equipments and medications for resuscitation should be readily available and checked at least daily in the delivery room and prior to any high risk delivery.

1.2 Delivery room should be warm and equipped with preheated radiant warmer. The infant should be dried thoroughly and quickly after birth.

1.3 Resuscitation should be started immediately when indicated, and not delayed for the assessment of one minute Apgar score.

1.4 Apgar score of ≤ 3 at 1 minute or ≤ 5 at 5 minutes will always require resuscitation and usually reflect severe hypoxic insult and acidosis.

1.5 If the 5-minute Apgar score is less than 7, additional scores are obtained every 5 minutes for a total of 20 minutes.

1.6 During resuscitation, the neonate should be placed on his back in slight Trendelenburg position with neck in a neutral position. A thin towel may be placed under the neonate's shoulders to maintain proper head position.

1.7 The mouth and nose should be suctioned using pressure not exceeding 136 cm H_2O (100 mm Hg). Avoid excessively vigorous suctioning which may cause reflex bradycardia and apnoea.

1.8 Drying and suctioning produce enough stimulation to induce effective breathing in most infants. Other stimulations are slapping the soles of the feet and rubbing the baby's back.

1.9 Infants in terminal apnoea (evidenced by slow [< 100/min] and falling heart rate, poor peripheral circulation) should be artificially ventilated immediately. Attempts to stimulate respiration by physical means in these infants invariably fail.

2. Meconium staining of amniotic fluid

2.1 Obstetrician should clear infant's oro- and nasopharynx by suctioning prior to delivery of shoulder.

2.2 In baby with thick meconium stained liquor, examine the laryngopharynx and cords with a laryngoscope immediately after delivery and clear the naso- and oropharynx by suctioning before intubation.

2.3 Intubate the baby and apply suction directly to the endotracheal tube while the tube is being slowly withdrawn. Repeat suctioning after reintubation until the trachea is clear of meconium.

2.4 In newborns without meconium at the level of vocal cords and who are clinically well, intubation may not be necessary.

3. Artificial ventilation

3.1 **Indications:**
  ❏ Any infant who fails to establish effective respiration following a brief period of stimulation.
  ❏ Apnoeic babies with a slow (< 100/min) and falling heart rate and poor peripheral circulation requires immediate artificial ventilation.

3.2 Ventilate at about 40–60 breaths/minute with bag and mask and 100% O_2.

3.3 Initial lung inflation may require pressures as high as 30–40 cm H_2O.

3.4 Assess adequacy of ventilation by observing the chest movement and by auscultation; adjust mask or the position of head if needed.

3.5 If ventilation is not satisfactory or likely to be prolonged, the baby should be intubated. The procedure of intubation should not take more than 30–45 seconds.

3.6 Severely asphyxiated infants and preterm infants: After spontaneous respiration has improved, it is better to wean off artificial ventilation gradually than to discontinue abruptly.

4. External cardiac massage

4.1 **Indications:** Heart rate < 60–80/minute with poor or absent peripheral pulse despite artificial ventilation.

4.2 **Technique:** The thumbs should be positioned on the sternum just below a line drawn between the nipples. The sternum is compressed ½ to ¾ inch at a rate of 120/minute. Compressions should be smooth, not jerky and equal in time to the relaxation phase.

5. Medications

5.1 **Naloxone:** to be given if the respiratory drive is thought to be depressed by narcotics given to the mother prior to delivery.
Dose: 0.1 mg/kg IV or via ET tube, may be repeated every 2–3 minutes as needed.

5.2 **Adrenaline:** produces vasoconstriction and improve cardiac contractility.
Dose: 0.01–0.03 ml/kg (0.1–0.3 ml/kg of 1:10,000 solution), may be repeated every 5 minutes if required by IV or via ET tube (dilute with 1–2 ml normal saline if given via ET tube).

5.3 **Sodium bicarbonate:** useful only in prolonged resuscitation. The baby should be monitored for blood gas and adequately ventilated for the elimination of CO_2 produced by sodium bicarbonate infusion.
Dose: 1–2 mEq/kg diluted to 0.5 mEq/ml with H_2O and infuse over 3–5 minutes to avoid transient hyperosmolarity.

6. Volume expanders

Indicated in the presence of hypovolaemia.
Dose: 10 ml/kg of plasma, normal saline or O-negative blood that has been crossmatched against mother's blood.

7. Factors for failure

Consider the followings if the baby does not respond to resuscitation:
7.1 Endotracheal tube in the wrong place — oesophagus or right main or lower lobe bronchus.
7.2 Inadequate inflation pressure.
7.3 Too small an endotracheal tube with large air-leak.
7.4 Oxygen has been disconnected.
7.5 Pneumothorax.
7.6 Anaemia.

8. Post-resuscitation care

Post-resuscitation care includes close monitoring of blood gases, blood calcium and glucose, correction of metabolic acidosis, treatment of hypotension, seizure, etc.

References

1. *Standards and guidelines for cardiopulmonary resuscitation and emergency cardiac care.* Part V: Neonatal Advanced Life Support JAMA 1986; 255:2969–73.

6

Approach to Cyanosis

LEE Hon-kwan

1. General approach

1.1 Distinguish peripheral cyanosis from central cyanosis. Infants with peripheral cyanosis have pink mucous membranes and normal PaO_2.

1.2 Distinguish intermittent cyanosis from persistent cyanosis. Common causes of intermittent cyanosis include apnoea and subtle seizures (Refer to Chs. 7 and 15 for details.)

2. Causes of persistent central cyanosis

2.1 Polycythaemia

2.2 Pulmonary
- ❏ Respiratory distress syndrome
- ❏ Pneumonia
- ❏ Meconium aspiration syndrome
- ❏ Diaphragmatic hernia
- ❏ Pneumothorax

2.3 Cyanotic congenital heart disease
- ❏ Transposition of great vessels
- ❏ Pulmonary atresia
- ❏ Critical pulmonary stenosis
- ❏ Tricuspid atresia

2.4 Persistent pulmonary hypertension of newborn

2.5 Others
- ❏ Methaemoglobinaemia
- ❏ Shock/sepsis

3. Approach to the diagnosis of persistent central cyanosis

3.1 Rule out the simple causes such as polycythaemia first.
  - ❏ History
    - Delayed clamping of cord
    - Twin delivery with twin-twin transfusion
    - Infant of diabetic mother
    - Small for date baby
  - ❏ Physical examination may show CNS depression, respiratory distress, heart failure and hypoglycaemia.
  - ❏ Diagnostic test: venous haematocrit > 65%.
  - ❏ Management: partial exchange transfusion with plasma.

3.2 Find out whether the cyanosis is due to respiratory diseases or cyanotic congenital heart disease.*
  - ❏ History
    - Prematurity is suggestive of RDS.
    - Prolonged rupture of membrane, maternal fever or foul-smelling liquor is suggestive of pneumonia.
    - Infant of diabetic mother has higher incidence of cyanotic congenital heart disease and RDS.
  - ❏ Physical examination
    The following is more suggestive of a cardiac cause:
    - Cyanosis with minimal respiratory distress
    - Single second heart sound
    - Presence of heart murmur
  - ❏ Investigations
    - Arterial blood gas — a high $PaCO_2$ is more suggestive of a pulmonary cause.
    - CXR — usually will show up the underlying pulmonary cause if it is present. The presence of cardiomegaly usually suggests a cardiac cause.
    - ECG
    - Hyperoxia Test — gives 100% of O_2 for 15 min. Take a pre-ductal arterial blood sample. PaO_2 of > 150 mm Hg makes cyanotic heart disease unlikely.
    - Echocardiography

3.3 Persistent fetal circulation (PFC).**
  - ❏ Suggestive clues

---

* For diagnosis and management of specific conditions, refer to Ch. 4 for pulmonary and Ch. 8 for cardiac conditions.
** For details of management, refer to Ch. 10.

- The presence of risk factors like perinatal hypoxia, hypo-glycaemia, post-maturity of polycythaemia.
- Loud pulmonary component of second heart sound.
- ECG shows R sided dominance
- CRX shows oligaemic lung field and no cardiomegaly.

❏ Diagnostic investigations
- Difference between pre-ductal and post-ductal PaO_2 > 15 mm Hg.
- Hyperventilation Test — improvement of PaO_2 when the $PaCO_2$ is driven down to a critical level (20–30 mm Hg).
- Contrast echocardiography — showing right-to-left shunt at atrial and/or ductal level.

3.4 Other causes

❏ Methaemoglobinaemia
Diagnosis — blood spot on blotting paper does not turn pink on exposure to air or oxygen.

❏ Shock/sepsis — usually other symptoms and signs and not cyanosis alone will dominate the clinical picture.

7

Apnoea

Alex KH CHAN

1. Definition

Cessation of spontaneous breathing for more than 10 sec associated with bradycardia of less than 100/min or cyanosis.

2. Causes

2.1 Apnoea of prematurity
2.2 Hyperthermia, hypothermia
2.3 Sepsis neonatorum
2.4 Infection of the Central nervous system: meningitis, encephalitis
2.5 Metabolic disturbances: Hypoglycaemia, hypocalcaemia, hyponatraemia.
2.6 Hypoxia
2.7 Severe respiratory distress
2.8 Anaemia
2.9 Upper airway obstruction
2.10 Excessive handling of the infant
2.11 Convulsion
2.12 Intracranial haemorrhage
2.13 Effect of transplacental transfer or postnatal adminstration of drugs.

3. Investigations

3.1 Dextrostix, followed by true blood glucose
3.2 Haemoglobin, white blood cell count, platelet count
3.3 Blood electrolytes including sodium, potassium, calcium and magnesium

3.4 Blood, urine and CSF for bacterial culture
3.5 Blood gases
3.6 X-ray chest
3.7 Ultrasound of the brain

4. Initial treatment

4.1 Physical stimulation e.g. flicking the sole of the feet.
4.2 Ventilate with face mask after clearing of airway if patient does not respond to physical stimulation.
4.3 Ventilate via endotracheal tube if the above fails.

5. Subsequent management

5.1 Correct hypoglycaemia, hypoxaemia, acid-base, electrolyte disturbances, anaemia, and temperature derangement if present.
5.2 Treat infection with appropriate antibiotics.
5.3 Intermittent Positive Pressure Ventilation for apnoeic patients suffering from severe respiratory distress.
5.4 Avoid excessive unnecessary handling, deep and prolonged pharyngeal suctioning, and hyperflexion of the neck.
5.5 Keep a clear upper airway, remove secretions.
5.6 Withhold feeding and give IV therapy if apnoea attack is related to feeding or regurgitation of feeds.
5.7 Continuous monitoring of heart rate, respiratory rate, oxygen saturation and blood pressure. Set respiratory delay alarm at 15 seconds and the heart rate alarm at 100/minute.

6. Treatment of apnoea of prematurity

6.1 Theophylline/aminophylline loading dose: 6 mg/kg oral/IV.
Maintenance dose: 2 mg/kg q12.
Effective plasma level: 6–12 mg/litre.
Note: Maintenance dose of aminophylline is best given by continuous infusion.
6.2 Caffeine citrate: loading dose: 20 mg/kg oral.
Maintenance dose: 5 mg/kg/day.
Therapeutic plasma level: 5–20 mg/litre.
Note: dosage should be halved when caffeine is used.
6.3 Doxapram: IV infusion at 2.5 mg/kg/hour.
Note: This drug can cause severe hypertension.

6.4 Continuous positive airway pressure (CPAP) may be used.
Use low pressure CPAP (3–4 cm H_2O).

6.5 Intermittent positive pressure ventilation is used when the above measures fail and patient has frequent/prolonged apnoea.

References

1. Marchal E, Bairam A, Vert P. *Neonatal apnoea and apnoeic syndromes.* Clin Perinatol 1987; 14(3):441–56.

2. Bairam A, Boutroy MJ, Badonnel Y et al. *Theophylline versus caffeine: Comparative effects in the treatment of idiopathic apnoea in the preterm infant.* J Pediatr 1987; 110:636–39.

3. Sagi E, Eya F, Alpan G, Patz D, Arad I. *Idiopathic apnoea of prematurity treated with doxapram and aminophylline.* Arch Dis Child 1984; 59(3):281–83.

4. Menon AP, Shefft GL, Thach BT. *Apnea associated with regurgitation in infants.* J Pediatr 1985; 106(4):625–29.

5. Miller MJ, Carlo WA, Martin RJ. *Continuous positive airway pressure selectively reduces obstructive apnea in preterm infants.* J Pediatr 1985; 106(1):91–94.

6. Henderson-Smart DJ, Butcher-Puech MC, Edwards DA. *Incidence and mechanism of bradycardia during apnoea in preterm infant.* Arch Dis Child 1986; 61(3):227–32.

7. Periman JM, Volpe JJ. *Episodes of apnea and bradycardia in the preterm newborn: Impact on cerebral circulation.* Pediatrics 1985; 76:333–38.

8. Hayakawa F, Hakamada S, Kuno D et al. *Doxapram in the treatment of idiopathic apnea of prematurity: Desirable dosage and serum concentrations.* J Pediatr 1986; 109:138–40.

9. Roberton NRC. *Manual of Neonatal Intensive Care.* 2nd ed. Edward Arnold Ltd., 1986; pp. 135–39.

8
Ventilator Therapy

FOK Tai-fai
William WONG

1. Indications*

Principally for RDS, but also applicable to most other neonatal pulmonary conditions.

1.1 Hypercapnia — $PaCO_2$ > 60 mm Hg especially when causing respiratory acidosis (pH < 7.20–7.25)

1.2 Severe hypoxaemia — PaO_2 < 50–60 mm Hg despite high FiO_2 (70–100%)

1.3 Severe retraction

1.4 Apnoea complicating course of RDS

2. Endotracheal tube

2.1 Size

Weight of infant (g)	Size (mm I.D.)
< 1000	2.5
1000–2000	3.0
2000–3500	3.5
> 3500	3.5–4.0

---

* In general, the smaller the infants, the more likely they require IPPV.
  Elective intubation and ventilation of VLBW (< 1500 g) infants has been reported to be associated with improved survival — a controversy.

2.2 Positioning

Body wt. (g)	Lip to tip distance (oral intubation)	Nare to tip distance (nasal intubation)
1000	7 cm	8 cm
2000	8 cm	10 cm
3000	9 cm	12 cm

2.3 ET Tube suctioning

❏ 4-hourly within first 24 hours; 2-hourly or hourly when secretions increase.

❏ Catheter size:
2.5/3.0 mm I.D. ET tube: No. 5 or 6 Fr
3.5 mm I.D. or larger: No. 8 Fr

❏ Instill 0.3 ml sterile saline before suction.

❏ Ventilate patient with 10% higher FiO_2 for 30 sec prior to and 30 sec after suction (or when pulse oximetry returns to pre-suction level).

❏ Duration of suction should not exceed 10 sec.

3. IPPV settings

Assuming a time-cycled, continuous flow, pressure-limited neonatal ventilator with IMV mode is being used.

3.1 Initial settings

	Infants with normal lung (e.g. those with apnoea)	Infants with RDS
Flow rate	~ 6 L/min	~ 6 L/min
PIP*	12–18 cm H_2O*	18–25 cm H_2O*
PEEP	2–3 cm H_2O	3–5 cm H_2O
Frequency	10–20 bpm, allow spontaneous resp.	40–60 bpm
IE ratio	1:2 to 1:10	1:1 to 1:2
Insp. time	0.5 sec	0.5 sec

* Use minimum PIP that can achieve satisfactory chest expansion and $PaCO_2$.

3.2 Subsequent adjustments

❏ Aim at maintaining:
PaO_2: 50–80 mm Hg
$PaCO_2$: 35–50 mm Hg
pH: 7.30–7.45

❏ Adjustments should be guided by blood gases which are affected by IPPV settings as follows:

IPPV setting	PaO_2	$PaCO_2$	Effect on pulm function	Adverse effects
↑ PIP	↑	↓	↑ tidal vol. ↑ MAP	barotrauma, airleaks BPD ↓ cardiac output
↑ PEEP	↑	↑	↑ FRC ↑ MAP	↓ tidal vol. ↓ lung compliance (when PEEP > 6 cm H_2O) ↓ cardiac output
↑ Frequency	±↑	↓	↑ minute ventilation	↓ insp. time → ↓ tidal volume ↓ exp. time → inadvertent PEEP ↓ lung compliance ↑ $PaCO_2$
↑ IE ratio	↑	±↑	↑ MAP	Airleak ↓ cardiac output ↓ exp. time → inadvertent PEEP ↑ $PaCO_2$
↑ FiO_2	↑	—	—	—
↑ Flow	±↑	±↓	↑ MAP	Inadvertent PEEP

❑ Points to note while adjusting ventilator settings:
  • Adjust one parameter per step (± adjusting FiO_2).
  • Check blood gas 15 minutes after each adjustment, less frequently if $TcPO_2$ (or SaO_2) and $TcPCO_2$ being continuously monitored.
  • Avoid PIP > 30 cm H_2O and PEEP > 6 cm H_2O.
  • A minimum PEEP of 2–3 cm H_2O is required by all intubated infants to maintain a satisfactory functional residual capacity.

4. Sudden deterioration during IPPV

Possible causes	Actions
Blocked ET tube	Change ET tube
Dislodged ET tube	Change ET tube
Tension pneumothorax	Transillumination Urgent XCR Diagnostic pleural tap if desperate (which is often the case)
Equipment failure	Check: power connection, circuit, leakage, gas source

5. Weaning from ventilator

5.1 Sequence of actions (example only)
❑ Reduce PIP to ≤ 25 cm H_2O.
❑ Reduce FiO_2 to 60%.
❑ Further reduce PIP to ≤ 20 cm H_2O.
❑ Reduce FiO_2 to ≤ 40%.
❑ Further reduce PIP/Reduce frequency as tolerated (guided by $PaCO_2$).
❑ Maintain a constant inspiratory time (e.g. 0.5 sec) while reducing IMV frequency.
❑ Infants may be disconnected from the ventilator at an IMV rate of 1–5 bpm, or weaned off IPPV through CPAP for infants weighing > 1000g.

5.2 Extubation
❑ Empty stomach.
❑ Clear nostrils, mouth, naso- and oropharynx.
❑ Extubate under direct vision using a laryngoscope, make sure the tracheal opening is not blocked by mucus.
❑ Administer O_2 (FiO_2 10% higher than that for IPPV) through oxygen headbox, subsequently adjust FiO_2 as guided by infants' PaO_2, $TcPO_2$ or SaO_2.
❑ Humidify oxygen headbox using an ultrasonic nebulizer for 24–48 hours after extubation.
❑ CXR 2 hours post-extubation to check for post-extubation atelectasis.

6. IPPV for other pulmonary conditions

6.1 Meconium aspiration syndrome
❑ Airleaks common, hence avoid high PIP or PEEP.
❑ Airway obstruction present, hence need sufficiently long expiratory time (I:E ratio of 1:2 or 1:3).

6.2 Pulmonary interstitial emphysema
❑ Try to reduce PIP, PEEP and MAP and tolerate a higher $PaCO_2$.
❑ If only one lung affected:
  • Position the infant such that the affected lung is dependent.
  • Selective ventilation of the unaffected lung and "resting" the affected lung may also be tried.
❑ High frequency oscillatory ventilation may be considered.

6.3 Persistent foetal circulation
❑ Paralyse infant with pancuronium/d-tubocurare.
❑ Maintain blood gases at:
  High range PaO_2 (e.g. 80 mm Hg)

PaCO$_2$: 25–30 mm Hg
pH: 7.5–7.55
❏ Often require high PIP, low PEEP (< 5 cm H$_2$O), and high frequency IPPV (~ 60 bpm) to achieve satisfactory blood gases.

6.4 Bronchopulmonary dysplasia
❏ Minimum PIP and FiO$_2$ to keep:
PaO$_2$: 50–70 mm Hg (or SaO$_2$ 88–92%)
PaCO$_2$: at high normal range

7. High frequency oscillatory ventilation (HFOV)

7.1 Indications
❏ Pulmonary interstitial emphysema (PIE).
❏ Pneumothorax treated with chest drain which is actively evacuating air.
❏ When conventional ventilation fails.

7.2 Settings (based on Infant Star neonatal ventilator)
❏ Starting HFOV on infants receiving IPPV/IMV:
• Monitor patient's oxygenation and ventilation continuously (TcPO$_2$, TcPCO$_2$, SaO$_2$).
• Before switching on HFOV, set its frequency at 10 Hz and amplitude at minimum.
• Switch HFOV on.
• Increase amplitude till infant's chest visibly vibrates.
• Take note of MAP displayed on ventilator.
• Reduce IPPV rate by half.
• Increase PEEP to maintain MAP at same level as displayed in ventilator.
• If PaCO$_2$ / TcPCO$_2$ continues to drop, reduce IMV rate (2–5 bpm/step) to 10 bpm.
• Do not change PIP unless PaO$_2$ falls or improves dramatically.
❏ Subsequent adjustments — to be guided by blood gases.
• ↑ PaO$_2$ by: ↑ FiO$_2$
                    ↑ MAP — by increasing PEEP
                    ↑ PIP
                    ↑ IPPV inspiratory time
                    ↑ IPPV rate
• ↓ PaCO$_2$ by: ↑ HFOV amplitude
                    ↑ HFOV frequency (most effective frequency is 10–22 Hz)
                    ↓ MAP (by decreasing PEEP) if lungs hyperinflated

7.3 Weaning from HFOV

❑ Ready when IMV rate < 10 bpm, MAP < 10 cm H_2O, and FiO_2 < 40%.
❑ Steps:
   • Do not alter frequency.
   • Reduce amplitude by steps of 2–5, stop reducing when $PaCO_2$ rises to your pre-determined limit.
   • Then reduce PEEP stepwise to what would be normal for IMV.
   • If $PaCO_2$ does not exceed your pre-determined limit, continue to reduce amplitude until it reaches its minimum, then turn off HFOV.
   • IMV rate may need to be slightly increased during weaning especially in apnoeic infants.

7.4 Precautions

❑ Must continuously monitor infant's oxygenation and ventilation.
❑ Back up IMV useful to prevent atelectasis.
❑ PEEP must be always ≥ 3 cm H_2O.
❑ Humidifier water must be at "full" level at all time to avoid amplitude drift.
❑ Secretions increase significantly at onset of HFOV; frequent tracheal suctioning essential especially during the first few hours of starting HFOV (often requires suctioning once every 15–30 minutes).
❑ CO_2 retention may be due to air trapping. Do not blindly increase amplitude to treat a rising $PaCO_2$. Rule out hyperinflation of lungs first (clinical ↑AP diameter of chest, CXR).
❑ Avoid over sedation as spontaneous breathing augments ventilation.
❑ Change back to conventional ventilation if HFOV not effective.

9
Respiratory Problems

LEE Wai-hong
KWONG Ngai-shan

A. HYALINE MEMBRANE DISEASE (IRDS)

1. Pathophysiology

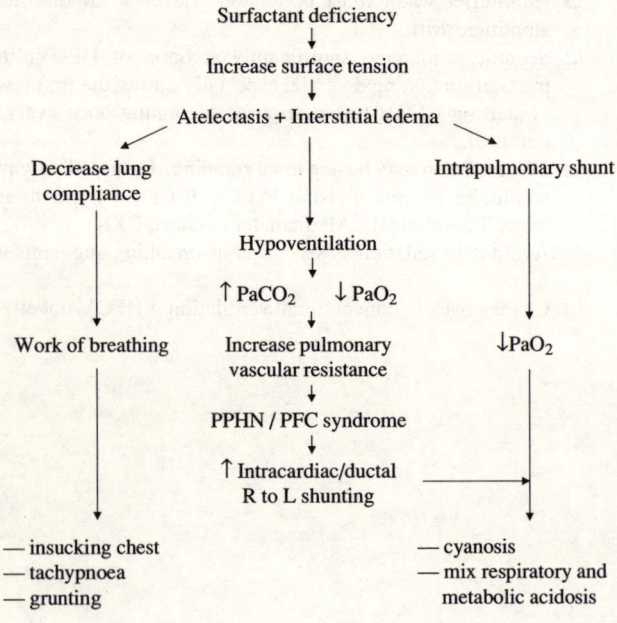

Surfactant deficiency

Increase surface tension

Atelectasis + Interstitial edema

Decrease lung compliance

Hypoventilation

$\uparrow PaCO_2$ $\downarrow PaO_2$

Intrapulmonary shunt

Work of breathing

Increase pulmonary vascular resistance

PPHN / PFC syndrome

$\uparrow$ Intracardiac/ductal R to L shunting

$\downarrow PaO_2$

— insucking chest
— tachypnoea
— grunting

— cyanosis
— mix respiratory and metabolic acidosis

2. Secondary effects

2.1 Metabolic and respiratory acidosis
2.2 Persistant pulmonary hypertension leading to increase R to L shunt at both intracardiac (via PFO) and transductal levels, i.e. persistant fetal circulation (PFC)
2.3 Multisystem hypoxaemic damages, resulting in
❑ myocardial depression with hypotension, tissue hypoperfusion, further aggravating metabolic acidosis;
❑ renal failure;
❑ gut ischaemia, predisposing to necrotising enterocolitis (NEC);
❑ CNS ischaemia, resulting in convulsion, etc.

3. Etiology

3.1 Prematurity (most important factor)
❑ esp. with gestation age less than 32 weeks
3.2 Perinatal asphyxia
3.3 Maternal diabetes mellitus (esp. poorly controlled)
❑ due to fetal hyperinsulinaemia
3.4 Caesarean section delivery
3.5 Sex (male to famale = 2:1)

4. Clinical features

4.1 Tachypnoea (respiratory rate > 60/minute)
4.2 Sternal retraction, intercostal and subcostal recession
4.3 Expiratory grunting — due to air flow through partially closed glottis in trying to prevent atelectasis
4.4 +/– cyanosis (not a constant feature)

Note: "Signs should be present *before* 4 hours of age, should still be there *at* 4 hours of age, and should persist for some period *beyond* 4 hours of age."
—N.R.C. Roberton

5. Radiological findings

5.1 Reticulogranular pattern (due to diffuse atelectasis)
5.2 Air bronchogram (air-filled major airways stand out as radiolucent areas)
5.3 Obliteration of cardiac border
5.4 Total white-out lung field

Note: Severity of HMD increases from 1 to 4.

6. Management

6.1 In labour ward
❏ Avoid perinatal asphyxia
  • Paediatrician standby for delivery
  • Immediate cardiopulmonary resuscitation for baby born with low Apgar scores (below 4), either bagging with face mask, or intubate and then manually ventilate, esp. babies with apnoea/irregular gasping
❏ Stabilize the baby, avoid cold stress:
  • Do not let baby exposed naked in wet.
  • Wrap up baby with warm towels after drying him up.
  • House baby in incubator.
❏ Observe baby closely. If cyanosis detected, give O_2 supplement (monitor SaO_2).
❏ Transfer to special/intensive care baby unit.

6.2 Subsequent management
❏ Place baby in incubator, set temperature to neutral thermal environment.
❏ Close monitoring of heart rate, respiratory rate.
❏ Keep SaO_2 / TcO_2 monitoring while on oxygen supplement.
❏ Monitor blood pressure if condition critical (may need to set up arterial line).
❏ Withhold feeding until symptoms improve.
❏ Regular monitoring of arterial blood gases.

7. Specific treatment

7.1 Oxygen therapy
❏ Use warm, humidified oxygen.
❏ Aim at:
  PaO_2 = 50–80 mm Hg / SaO_2 = 90–95%
  $PaCO_2$ = 35–45 mm Hg / pH = 7.35–7.45
❏ Can be given to incubator or headbox if symptoms mild.
❏ Consider pressure support as well if O_2 required more than 0.6 to keep satisfactory PaO_2.

7.2 Pressure support
❏ Aim to distend the alveoli (prevent atelectasis, keep FRC):
  (a) Continuous Positive Airway Pressure (CPAP):
    • For moderate HMD when PaO_2 < 60 mm Hg in 60% O_2

- Can be given via:
  nasopharyngeal
  endotracheal (not preferred)
  nasal
- Device used:
  endotracheal tube (ETT)
  nasal prongs
- Nasopharyngeal CPAP: pass ETT via the nostril until the tip is visualized just behind the soft palate; secure the tube in place by strappings.
- Initial CPAP settings:
  Pr = 4 to 6 cm water
  FiO_2 = 0.6 (adjust according to PaO_2/SaO_2)
  Flow = 8–10 L/min
  (Pressure can be stepped up to 8 cm with 2 cm increment if requirement to raise the PaO_2.)
- Stop feeding while baby is on CPAP. Stomach should be constantly emptied because of risk of aspiration pneumonia in case he regurgitates. However, for prolonged CPAP, feeding can be given via nasogastric/nasojejunal routes.
- CPAP may also be indicated during weaning off from respirator (see below) or recurrent apnoea.

(b) Intermittent Positive Pressure Ventilation (IPPV):
- Indications for IPPV in HMD:
  1. Sudden deterioration with apnoea or irregular gasping
  2. Respiratory failure, due to severe HMD, to establish satisfactory respiration after resuscitation in labour ward
  3. Deteriorating blood gases (despite CPAP)
- Requires intubation
- Usual initial settings used:
  FiO_2 at previous level or 10% higher
  PEEP 4 to 6 cm H_2O
  PIP 16 to 20 cm H_2O
  Rate 40–60/minute
  I/E ratio = 1:1 to 1.5:1
- Basic principles:
  1. Use lowest PIP/PEEP possible to avoid barotrauma. Best keep PIP below 20 cm if possible.
  2. Avoid using too high PEEP (> 8 cm H_2O) as it will impair cardiac output.
  3. Reduce the airway pressure (both PIP/PEEP) by small decrement as soon as blood gas improves.
  4. Avoid using 100% FiO_2 because of oxygen toxicity and the theoretical possibility of diffuse atelectasis after prolonged use.

Table 1. Usual effects of changing ventilator settings

Increasing	PaO$_2$	PaCO$_2$	pH	Complications
		Causes		
FiO$_2$	↑	0	0	Oxygen toxicity (bronchopulmonary dysplasia, retrolental fibroplasia); absorption atelectasis
CPAP/PEEP	↑	0 / ↑	0 / ↓	Hypoventilation with respiratory acidosis; decreased cardiac output with metabolic acidosis; air leaks
PIP	↑	↓	↑	Barotrauma with air leaks and bronchopulmonary dysplasia; respiratory alkalosis
Rate	↓	↓	↑	Respiratory alkalosis
I/E ratio (1:1 to 3:1)	↑	0	0	Increased intrapleural pressure; decreased venous return

(c) Weaning from ventilator:
- Should be considered when conditions improve:
  1. Arterial blood gases stable in physiological range
  2. Spontaneous respiratory effort
  3. Increase activity and muscle tone
  4. Progressive decreasing FiO$_2$ requirement
- Consider weaning from IPPV/IMV to CPAP/headbox oxygen when: PIP < 15 cm H$_2$O
  PEEP < 4 cm H$_2$O
  FiO$_2$ < 0.4
  Rate < 10/min (without apnoea)
- Exercise discretion in individual cases
- When change from IMV to CPAP:
  1. Use the previous PEEP as guideline, usu. about 5 cm water.
  2. Clear the ETT and oropharynx first; empty the stomach also.
  3. Then pull the ETT up for nasopharyngeal CPAP.
  4. Withhold feeding for several hours.
- When change from IPPV/CPAP to headbox:
  1. Clear the oropharynx and empty the stomach.
  2. Use current FiO$_2$ as guideline. Increase by 5–10% if required.
  3. Extubate and lie the baby prone.
  4. Withhold feeding for several hours.

- In both situations above:
1. Keep close monitoring of cardiopulmonary functions throughout.
2. Monitor for apnoea.
3. SaO_2/PaO_2, blood gases monitoring.
4. Take X-ray chest for possible atelectasis.

8. Other therapeutic agents

8.1 Antibiotics
❑ When infection cannot be confidently ruled out.

8.2 Sedation
❑ Usually for calming the baby while on IPPV, reduces "fighting" against ventilator, and to help better synchronisation.
❑ Commonly used: morphine 0.05 mg q6h IV.

8.3 Neuromuscular blockers
❑ Not routinely administered.
❑ Help reduce incidence of periventricular haemorrhage.
❑ Must be stopped completely during weaning from ventilator (either use in full dose or stop all).
❑ Drugs commonly used:
- Pancuronium (Pavulon) 0.5–0.5 mg/kg per dose IV prn
- Atracurium (Tracrium) 0.3–0.5 mg/kg per dose IV prn. Continuous infusion 0.5 mg/kg/hr
- Vecuronium (Norcuron) 0.1–0.2 mg/kg per dose.

9. Newer therapeutic approaches

9.1 Surfactant replacement therapy
❑ Types:
- Natural — from human amniotic fluid, cow's or pig's lung extract;
- Semi-synthetic — natural surfactants with modification
- Synthetic — ease of manufacturing; no foreign proteins
❑ Modes:
- Prophylactic
- Rescue
❑ Preparations:
- Exosurf (Wellcome)
  - Synthetic
  - Intratracheal instillation
  - Consists of dipalmitoylphosphatidylcholine (DPPC) for surface tension lowering;

Cetyl alcohol & tyloxapol (to facilitate rapid spreading and adsorption of DPPC throughout the alveolus/air interface) NaCl (adjust tonicity up to 0.1 N)
- Dosage: 5 cc/kg in 2 half-doses to right and left side; repeat in 12 hr if needed
- Survanta (Abbott)
  - Semisynthetic (from homogenized cow lungs with modifications)
  - Intratracheal instillation in 4 aliquots
  - Dosage (see dose chart from manufacturer)

9.2   High frequency ventilation (HFV)

❑ Using high frequency (up to 15–20 Hz), low tidal volume mode, hoping to reduce the barotrauma associated with conventional ventilation technique

❑ 2 commonly used HFV:
- HFJV (jet ventilation)
- HFOV (oscillatory ventilation)

References

1. Roberton NRC. *A Manual of Neonatal Intensive Care*. 2nd ed. London, Arnold, 1986; pp. 72–122.

2. Merenstein GB, Gardner SL. *Handbook of Neonatal Intensive Care*. 2nd ed. Philadelphia, Mosby, 1989; pp. 365–427.

3. Wetzell R, Gioia FR. *High frequency ventilation*. Pediatric Clinics of North America, 1987; 34(1):15–38.

4. Guthrie RD. *Neonatal Intensive Care*. 1st ed. New York, Churchill Livingstone, 1988; pp. 21–74.

5. Avery ME, First LR. *Paediatric Medicine*. 1st ed. Baltimore, Williams and Wilkins, 1989; pp. 165–69.

B. MECONIUM ASPIRATION SYNDROME

1. Pathophysiology

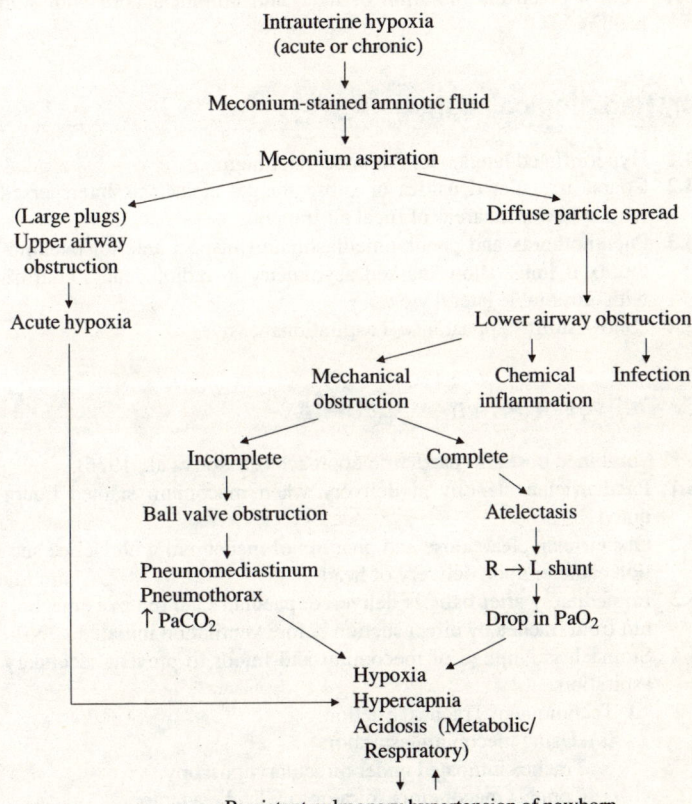

2. Diagnostic criteria

2.1 Meconium in oropharynx or trachea
2.2 Clinical evidence of respiratory distress
2.3 Radiological signs

3. Clinical recognition

3.1 Respiratory distress starts shortly after birth, worsening over subsequent 24 hours

3.2 Tachypnea and overdistended chest

3.3 Crepitations and rhonchi frequently present

3.4 Yellow-green discoloration of nails and umbilical cord with skin peeling

4. Radiological signs

4.1 Hyperinflated lungs with flattened diaphragm

4.2 Coarse irregular densities of subsegmental atelactasis interspersed with hyperlucent areas of focal air trapping

4.3 Pneumothorax and pneumomediastinum (Suspect anterior pneumothorax if lungs show marked asymmetry in radiolucency; confirm with cross-table lateral view.)

4.4 "Snow-storm" appearance if aspiration massive

5. Delivery room management

Combined obstetric-paediatric approach (Carson et al., 1976).

5.1 Paediatrician standby at delivery when meconium stained liquor noted.

5.2 Obstetrician clears nose and pharynx of meconium with DeLee suction catheter upon delivery of head.

5.3 Immediately after baby is delivered, paediatrician removes meconium from trachea by direct suction before ventilation initiated.

5.4 Stomach is emptied of meconium and liquor to prevent secondary aspiration.

❑ Technique of Tracheal Suction:

(a) Using meconium aspirator:

- Trachea intubated under direct laryngoscopy.
- Connect meconium aspirator to ET tube adaptor.
- Attach suction tube to meconium aspirator.
- Apply suction to trachea by closing thumb-control port while withdrawing ET tube.
- Quickly repeat until clear.
- Avoid positive pressure ventilation before tracheal suction.
- Keep watch on general condition of baby during suctioning: interrupt process and oxygenate promptly if significant bradycardia appears.

6. Post-delivery room management

6.1 Admit to Special Care Nursery if further monitoring required.

6.2 Monitoring and evaluation:
- ❑ Continuous cardiorespiratory monitoring.
- ❑ Blood pressure.
- ❑ Transcutaneous O_2 and CO_2.
- ❑ Chest X-ray.
- ❑ Blood gases: arterial catheter desirable if $FiO_2 > 30$–40% required.

6.3 Start antibiotics, usually ampicillin plus an aminoglycoside (meconium enhances bacterial growth).

6.4 Give oxygen to keep PaO_2 around 80–100 mm Hg.

6.5 Ventilate if respiratory failure develops.

6.6 Start tolazoline therapy if persistent pulmonary hypertension causes hypoxaemia refractory to ventilator therapy.

6.7 Avoid chest physiotherapy in acute phase: PaO_2 can drop precipitously especially if PFC present. If necessary, suction trachea again carefully after pre-oxygenation.

6.8 Attend to associated complications of perinatal anoxia, e.g., cerebral oedema, seizures, etc.

6.9 For babies not admitted for special care:
- ❑ Take CXR and monitor vital signs for first day.
- ❑ Preferably start antibiotics, taking off after 48 hours if remaining asymptomatic and CXR normal.

6.10 Look out for Listeriosis if meconium staining occurs with preterm delivery.

7. Ventilator strategy

7.1 Adjust FiO_2 to keep PaO_2 80–100 mm Hg.

7.2 Allow adequate expiratory time to avoid further air trapping.

7.3 Avoid high PIP/PEEP:
- ❑ Risk of air leaks.
- ❑ Pressure-splinting of lungs decreases compliance and worsens $PaCO_2$.
- ❑ Obstructs venous return causing BP drop.

7.4 Fast rates preferred to higher pressures in lowering $PaCO_2$.

7.5 Paralyse with pancuronium when:
- ❑ PIE appears.
- ❑ Baby restless on ventilator.
- ❑ High PIP/PEEP.

7.6 Try HFOV when PIE or pneumothorax developed.

Complications to anticipate

Acute deterioration — Tension pneumothorax
— ET tube blocked by meconium

Rising PaCO$_2$ — ET tube obstruction by meconium
PIE

Persistent hypoxaemia — Persistent pulmonary hypertension

Hypotension — Tension pneumothorax
— Excessive PIP/PEEP
— Myocardial dysfunction

References

1. Vidyasagar D, Yeh TF et al. *Assisted ventilation in infants with meconium aspiration.* Pediatr 1975; 56:208–13.
2. Gregory GA, Gooding CA et al. *Meconium aspiration in infants — a prospective study.* J. Pediatr 1974; 85:848–52.
3. Carson BS, Losey RW et al. *Combined obstetric and paediatric approach to prevent meconium aspiration syndrome.* Am. J. Obs. Gynaec 1976; 126:712–15.
4. Ting P, Brady J. *Tracheal suction in meconium aspiration.* Am. J. Obs Gynaec 1975; 122:767–71.
5. Eisner P. *Suctioning meconium from trachea. A new solution to an old problem.* Pediatr 1986; 713.
6. Yu VYH. *Respiratory Disorders in the Newborn.* Edinburgh, Churchill Livingstone, 1986; pp. 29–36.
7. Vyas H, Milner AD. *Other Respiratory Diseases in the Neonate.* In Roberton NRC (ed): *Textbook of Neonatology.* Edinburgh, Churchill Livingstone, 1986; pp. 317–19.

C. PULMONARY AIR LEAKS

1. Pathogenesis

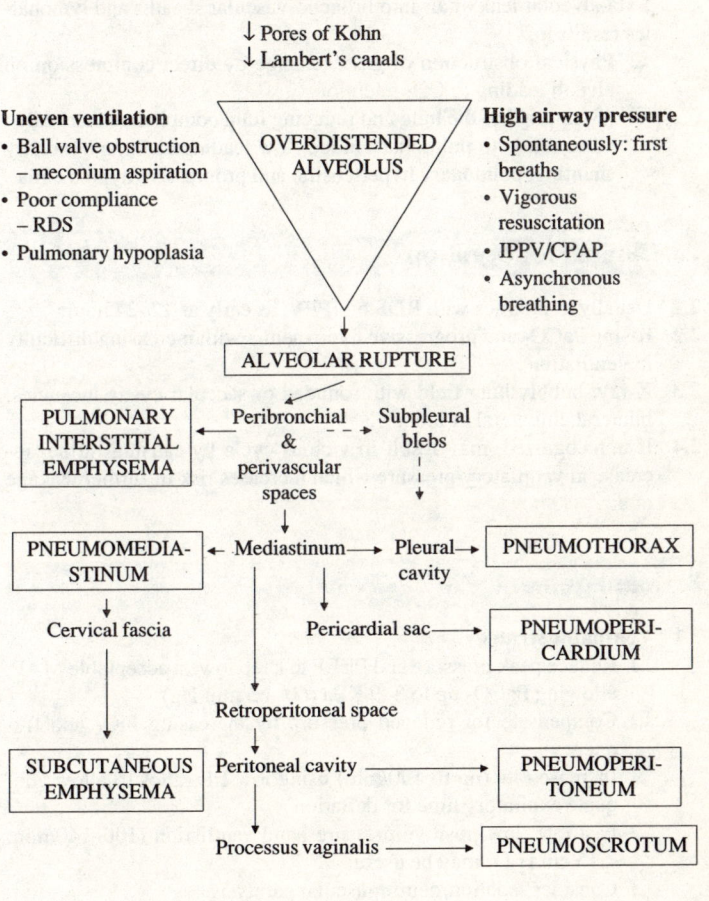

↓ Pores of Kohn
↓ Lambert's canals

Uneven ventilation
- Ball valve obstruction
  – meconium aspiration
- Poor compliance
  – RDS
- Pulmonary hypoplasia

OVERDISTENDED
ALVEOLUS

High airway pressure
- Spontaneously: first
  breaths
- Vigorous
  resuscitation
- IPPV/CPAP
- Asynchronous
  breathing

ALVEOLAR RUPTURE

PULMONARY
INTERSTITIAL
EMPHYSEMA

Peribronchial
&
perivascular
spaces

Subpleural
blebs

PNEUMOMEDIA-
STINUM

Mediastinum → Pleural
cavity

PNEUMOTHORAX

Cervical fascia

Pericardial sac

PNEUMOPERI-
CARDIUM

Retroperitoneal space

SUBCUTANEOUS
EMPHYSEMA

Peritoneal cavity

PNEUMOPERI-
TONEUM

Processus vaginalis

PNEUMOSCROTUM

C1. PULMONARY INTERSTITIAL EMPHYSEMA (PIE)

1. Pathophysiology

Exta-alveolar leak of air into broncho-vascular sheaths and lymphatics results in:
- ❏ Physical obstruction of gas exchange by direct compression on alveoli leading to CO_2 retention
- ❏ "Splinting" of the lung and reducing lung compliance markedly
- ❏ Obstruction to pulmonary blood flow leading to intrapulmonary shunting, pulmonary hypertension and progressive hypoxaemia

2. Clinical recognition

2.1 Usually in premies with RDS on IPPV, as early as 12–24 hours

2.2 Rising $PaCO_2$ and progressive hypoxaemia with increasing difficulty in ventilation

2.3 X-ray: bubbly lung field with rounded or saccular cystic lucencies: bilateral, unilateral or lobar

2.4 If unrecognized, may result in vicious cycle by causing further increase in ventilatory pressure which increases risk of further leakage of air.

3. Management

3.1 **Ventilator strategy**
- ❏ Reduce peak pressure and PEEP to keep lowest acceptable MAP, allowing $PaCO_2$ up to 8–9 KPa (60–65 mm Hg)
- ❏ Compensate for reduced pressure by increasing FiO_2 and frequency
- ❏ Increase rate (up to 150/min) using low I:E ratios to allow adequate expiratory time for deflation
- ❏ Fast rate, low positive pressure hand ventilation (100–140/min, < 15 cm H_2O) may be useful
- ❏ Consider sedation/neuromuscular paralysis
- ❏ Switch to High Frequency Oscillatory Ventilation if blood gases cannot be maintained on conventional ventilation

3.2 **Special manoeuvres for unilateral PIE**
- ❏ Place infant in decubitus position with affected lung in dependent position
- ❏ Rest affected lung:
  - Selective bronchial intubation of unaffected lung

- Selective bronchial obstruction of affected lung by balloon catheter
❑ Watch for pneumothorax

C2. PNEUMOTHORAX

1. Clinical recognition

1.1 Asymptomatic if small
1.2 Respiratory distress as primary presentation or as worsening of pre-existing respiratory symptoms
1.3 **Associated signs**
- ❑ Ipsilateral chest bulge and abdominal distension
- ❑ Diminished breath sounds, increased resonance
- ❑ Irritability
- ❑ Vital signs: transient $\uparrow$ BP, $\downarrow$ pulse pressure, $\downarrow$ $PaO_2/TcPO_2$, $\uparrow$ $PaCO_2/TcPCO_2$, reduced transthoracic impedance
1.4 **Signs of tension**
- ❑ Sudden marked deterioration
- ❑ Pallor/cyanosis/peripheral vasoconstriction
- ❑ Hypotension, bradycardia, apnoea
- ❑ Contralateral shift of cardiac impulse
1.5 **Pitfalls**
- ❑ Auscultation can be deceptive: bronchotubular breath sounds can be well transmitted across pneumothorax
- ❑ Mediastinal shift may not be appreciable if:
  - Bilateral tension pneumothoraces
  - PIE in contralateral lung
  - Poorly compliant lungs as in HMD: affected lung may fail to collapse and contralateral lung too stiff to permit appreciable shift

2. Transillumination

2.1 Positive if affected hemithorax lights up in translucent glow when fibreoptic cold light probe placed on chest wall.
Useful for immediate detection of life-threatening pneumothorax which requires urgent treatment before X-ray can be obtained, therefore examine with transilluminator whenever an at risk infant shows sudden deterioration.
2.2 **False positive**
- ❑ chest wall oedema

❏ Subcutaneous emphysema
❏ PIE
❏ Extremely small infant

2.3 False negative
❏ Small pneumothorax
❏ Bilateral pneumothoraces
❏ Thick chest wall

2.4 Baseline transillumination in at risk infants improves diagnostic accuracy.

3. Radiological signs

3.1 Classical sign (AP view) free lateral lung border separated from chest wall by radiolucent space devoid of lung markings.

3.2 Signs of tension
❏ Mediastinal displacement
❏ Herniation of pleural space across anterior and superior mediastinum
❏ Fattened diaphragm
❏ Bulging of intercostal space between ribs

3.3 Pitfalls
❏ Skin fold may mimic free lung border.
❏ In RDS babies, significant lung collapse often not seen because of poor lung compliance.
❏ Anterior pneumothorax presents as hyperlucent hemithorax without free lung edge visible in the suspine film.
❏ Medial pneumothorax presents as paramediastinal lucency or enhanced sharpness of mediastinal border (without outlining thymus).
❏ Subpulmonic collection presents as enhanced costophrenic sulcus or lucent diaphragmatic rim.

3.4 Supplementary views
❏ Horizontal beam lateral — useful in localizing site of air collection
❏ Lateral decubitus with suspected side up — may show up small pneumothorax

4. Measures to prevent air leaks during IPPV

4.1 Active monitoring of babies mandatory. Periodic chest transillumination for early detection of pneumothorax

4.2 Use lowest possible PIP & PEEP

4.3 Allow adequate expiratory time to minimize air trapping (especially with meconium aspiration, diaphragmatic herina, pulmonary hypo-plasia)

4.4 Eliminate asynchronous breathing by
- ❏ Increase ventilator frequency to above spontaneous rate.
- ❏ Shorten inspiratory time (< 0.8 sec).
- ❏ Consider sedation or paralysis if fighting persists despite correction of underventilation.
- ❏ Synchronizing mechanisms to synchronize ventilator cycle to baby's spontaneous effort has been developed.

4.5 Incorporate manometer in ventilation circuit to control inflating pressure during hand ventilation.

5. Conservative management

If:
- ❏ Pneumothorax small and not under tension
- ❏ Minimal distress
- ❏ Not on IPPV

6. Emergency chest tapping

6.1 Temporary evacuation of life threatening air accummulation while preparing for permanent tube placement

6.2 Indicated if
- ❏ Signs of tension pneumothorax
- ❏ Acute cardio-respiratory collapse (without waiting for X-ray confirmation if transillumination positive)

6.3 **Technique**
- ❏ Prepare skin with antiseptic.
- ❏ Insert angiocath in 2nd I.C.S. in mid-clavicular line just over top of rib.
- ❏ Once catheter tip entered pleura (marked by rapid outflow of gas), slide cannula into pleural cavity while withdrawing stylet.
- ❏ Attach 10 ml syringe with 3-way stopcock to cannula and gently aspirate air from pleural space.
- ❏ Continue evacuation while preparing for permanent tube placement; connect by tubing to temporary underwater seal if air-leak is continuous and baby's condition is unsatisfactory.

6.4 **Alternative**
- ❏ Use 21G–23G butterfly or needle connected via stopcock to syringe. Apply suction gently with syringe as needle/butterfly enters pleura. Avoid excessive depth of insertion to prevent lung laceration and remove needle as soon as possible.

7. Thoracostomy tube drainage

7.1 Indications
- ❏ Continual air leak
- ❏ Ongoing distress with coexisting pulmonary disease
- ❏ On IPPV
- ❏ After needle aspiration for tension pneumothorax

7.2 Drainage system
- ❏ Thoracostomy tube: 10 Fr or 12 Fr Argyle ± trocar
- ❏ Evacuation device:
  - • Single bottle drainage with regulated suction
  - • 2-bottle suction system
  - • Pleur-evac system (and related types), combining collection, water seal and suction chambers in one unit
  - • Heimlich one-way flutter valve useful during transport

Table 2. Performance of Chest Tube: Trouble Shooting

Behaviour of drain		X-ray	Diagnosis	Action
Continuous bubbling	Increase with more suction	Air ±	Leak in system	Remove leak
	Positional		Dislodged tube (or side-holes out of chest)	Replace/remove
	Increase with more suction Not positional Bloody aspirate Trocar used for insertion	Air +	Bronchopleural fistula	Off suction Await healing Consult surgeon if uncontrolled
	Clears with increased suction	Air +	Inadequate suction	Increase suction 5–10 cm
Intermittent bubbling	During ventilator inspiratory phase only	Air evacuated	Tube patent, pleural leak still present	Continue drainage until bubbling stops
	More bubbles in ventilator inspiratory phase	Air +	Tube patent, suction inadequate	↑ suction 5–10 cm Check equipment Insert 2nd tube PRN
			Air loculated tube not in contact	Reposition baby Insert 2nd tube PRN
	More bubbles in ventilator expiratory phase	Air +	Suction too high — tube occluded by lung	Reduce suction Rotate tube 1/4 turn

Table 2. Performance of Chest Tube: Trouble Shooting (continued)

Behaviour of drain		X-ray	Diagnosis	Action
No bubbling	Meniscus fluctuates with respiration when off suction	Air evacuated	Tube patent, no continuous leak	Remove tube
		Air +	Tube patent, air loculated	Reposition baby Gentle percussion Insert 2nd tube PRN
			Tube patent, extra-pleural air	Other X-ray views Consider direct aspiration
	Fluctuates with heart beat	Air ±	Tube patent, detecting cardiac impulse	Pull tube away from mediastinum
	No fluctuation: static meniscus	Air +	Tube occluded – fibrin/blood clot – external compression/kink – malposition (mediastinum, sub-cutaneous)	Strip; replace PRN Remove kink Reposition/replace tube
		Air evacuated	Tube occluded	Remove tube

* Modified from: MA Fletcher, MG MacDonald and GB Avery, *Atlas of Procedures in Neonatology* Lippincott JB, 1983.

Table 3. Other Types of Pulmonary Air-leaks

Type	Clinical recognition	Radiological sign	Significance	Treatment
Pneumomediastinum	No specific sign Transillumination equivocal	AP: – hyperlucent rim to lateral borders of mediastinum – uplifted thymic lobes — "Angel's Wing Sign" Horizontal beam lateral: – Air in anterior mediastinum	Likely to precede pneumothorax	Reduce ventilator pressures No specific drainage
Pneumopericardium	Substernal transillumination Signs of tamponade: – ↓ BP & pulse pressure – ↓ pulse volume tachycardia – ↓ ECG size – persistent pulsation of fluid in UAC	Air encircling heart and outlining pericardial sac	Look for other air-leaks Unrelieved cardiac tamponade fatal	Subxiphoid needle aspiration if tamponade ± pericardial drain Reduce ventilator pressure

Table 3. Other Types of Pulmonary Air-leaks (continued)

Type	Clinical recognition	Radiological sign	Significance	Treatment
Pneumoperitoneum	Tympanitic abdominal distention Abdominal transillumination	AP: – air under diaphragm – faliciform ligament sign (Football sign) – pneumo-scrotum Horizontal beam lateral: – free peritoneal gas	Differentiate from ruptured hollow viscus Tension effects if very large	Paracentesis if under tension and splinting diaphragm
Subcutaneous emphysema	Crepitus	Lacy lucencies in soft tissue	Look for other air-leaks	Not necessary
Systemic air embolism	Sudden collapse Air bubbles in blood drawn from arterial catheter	Gas shadow in heart chambers and vascular tree	Other serious air-leaks already present Extremely excessive pressure	None effective; invariably fatal

D. MASSIVE PULMONARY HAEMORRHAGE

1. Nature

Fulminant haemorrhagic pulmonary oedema with leakage of capillary filtrate and red cells into alveolar spaces (Cole et al., 1973).

2. Pathogenesis[1]

2.1 Increased net pulmonary capillary filtration pressure.
- ❏ Acute left ventricular failure causing raised pulmonary capillary pressure
- ❏ Low plasma oncotic pressure

2.2 Damage to alveolar membrane.

2.3 Coagulation defect probably prolong or exacerbate MPH rather than initiate it.

3. Predisposing circumstances[2]

Usually in premature/small for dates baby with:
- ❏ Hypoxia and acidosis
- ❏ Asphyxia
- ❏ Cold injury
- ❏ Acute fluid overload
- ❏ Congenital heart disease with increased pulmonary blood flow
- ❏ HMD with IPPV
- ❏ Severe sepsis
- ❏ After surfactant therapy

4. Clinical recognition

4.1 Most common during first 4 days

4.2 Acute sudden deterioration:
- ❏ Gasping respirations ending in apnoea (if not on IPPV)
- ❏ Cyanosis (hypoxaemia)
- ❏ Pallor, mottling (peripheral vasoconstriction)
- ❏ Bradycardia, hypotension
- ❏ Flaccid

4.3 Outpouring of fresh frothy blood-stained fluid from oropharynx or ET tube:

❑ PCV of fluid typically less than 10% (at least 15–20% less than circulating blood)[1,3]

❑ Volume varies from 2–10 ml

4.4 Acidosis, drop in PCV, signs of DIC may follow subsequently.

5. Radiological picture

❑ Fluffy densities, or

❑ Coarse nodular pattern, or

❑ Homogeneous opacification of both lung fields in severe cases

6. Differential diagnosis

6.1 Massive IVH can cause similar picture of deterioration.

6.2 Direct mucosal trauma to trachea and bronchi by suction catheter also yields bloody aspirate from ET tube.

❑ Differentiation: less copious, not foamy

6.3 Coagulopathy do not cause isolated pulmonary bleeding only.

7. Management

Immediate resuscitation is required:

7.1 Clear airway of blood.

7.2 Intubate and start IPPV with PEEP to reduce blood/fluid effux.[4]

7.3 Raise PIP and PEEP if already on IPPV.

7.4 Start Dopamine infusion for inotropic support.

7.5 Control LV failure with diuretics and morphine if acute fluid overload.

7.6 Control fluid and blood administration carefully: baby usually does not have large volume loss, while excessive volume replacement will exacerbate the pulmonary oedema.[3]

7.7 Slow blood transfusion (consider packed cells) or exchange transfusion to correct blood loss/anaemia.

7.8 Evaluate for coagulopathy: give vit. K, platelets, FFP as appropriate.

References

1. Cole V, Normand ICS, Reynolds, EOR et al. *Pathogenesis of haemorrhagia pulmonary oedema and massive pulmonary haemorrhage in the NB*. Peds 1973; 51:175–87.

2. Rowe S, Avery ME. *Massive pulmonary haemorrhage in the newborn II. Clinical considerations.* J Pediatr 1966; 69:12–20.

3. Hansen T, Corbet A. *Pulmonary Haemorrhage.* In Taeusch HW, Ballard RA, Avery ME (eds). *Diseases of the Newborn.* 6th ed. Philadelphia, WB Saunders, 1991; pp. 485–86.

4. Trompeter R, Yu VYH, Aynsley-Green, Roberton NRC. *Massive pulmonary haemorrhage in the newborn infant.* Arch Dis Child 1975; 50:123–27.

E. BRONCHOPULMONARY DYSPLASIA (BPD)

1. Diagnostic criteria[1]

1.1 IPPV during first week for minimum of 3 days

1.2 Chronic respiratory distress: tachypnoea, retractions and rales, persisting beyond 28 days

1.3 Oxygen requirement beyond 28 days to maintain PaO_2 over 50 mm Hg

1.4 Chest X-ray showing persistent strands of radiodensity alternating with areas of increased lucency

2. Staging of BPD[2]

Stage		Age (days)	Radiologic features
I	Acute RDS	1–3	Granular pattern
II	Regeneration	4–10	White-out
III	Transition to chronic disease	10–20	Small bubbly lucencies alternating with areas of irregular densities
IV	Chronic disease	> 30	Large heterogeneous areas of emphysema alternating with coarse infiltrates

Note: The picture more often seen is characterized by:
1. insidious progression of chronic global involvement
2. spectrum of variable hyperaeration with a fine, lacy, ill-defined pattern of interstitial densities, usually noticeable by 3rd–4th week
3. occasionally preceeded by hazy appearance of lung fields

3. Differential diagnosis

3.1 Clinical: causes of chronic respiratory distress
- ❏ Chronic lung disease (CLD):
  - • Bronchopulmonary Dysplasia (BPD)
  - • Chronic Pulmonary Insufficiency of Prematurity (CPIP)
  - • Wilson-Mikity Syndrome
- ❏ Perinatally acquired CMV pneumonia
- ❏ Chronic aspiration syndromes
- ❏ Pulmonary hypoplasia
- ❏ Upper airways/tracheal obstruction
- ❏ Rickets of prematurity
- ❏ Congestive heart failure
- ❏ Neuromuscular weakness
- ❏ Immaturity of respiratory control

3.2 Radiographic DDX[3]
- ❏ In clinical setting comparable to that of BPD:
  - • PIE
  - • Acquired CMV pneumonia
  - • Wilson-Mikity syndrome
  - • Fluid overload (uncommon)
- ❏ In clinical setting different from BPD:
  - • Meconium aspiration syndrome
  - • Pulmonary lymphangiectasia
  - • TAPVD with obstruction
  - • Recurrent pneumonia (GER, H-type tracheo-oesophageal fistula)
  - • Cystic fibrosis
  - • Idiopathic pulmonary fibrosis

4. Management

4.1 **Objectives**
- ❏ To minimize further lung damage
- ❏ To enhance growth and repair of lungs

4.2 **Respiratory care**
- ❏ Ventilator therapy
  (see chapter 5)
- ❏ Blood gases monitoring
  Continuous tracking of $TcPO_2/TcPCO_2$ or pulse oximetry with intermittent blood gases measurements.
- ❏ Airway care
  - • scrupulous ET tube care to prevent problems of prolonged intubation

- physiotherapy to remove secretions and prevent atelactasis
❑ Wearning strategy
  - Adopt sequence in Chapter 5, but in small decrements, allowing one to several days before the next step depending on tolerance.
❑ Oxygen therapy
  - Supplementary oxygen often required for weeks to months after extubation
  - Watch for signs of silent hypoxia:
    - Insiduous increase in respiratory rate
    - Stall in weight gain
    - Rise in pulmonary arterial pressure
  - Wean oxygen slowly by small decrements in flow
  - Change to nasal cannula for O_2 delivery from low-flow flowmeter when:
    - $FiO_2 < 0.3$
    - No labile changes with routine nursing care
    - Growing
  - Prevent hypoxia during periods of stressful activity (feeding, physiotherapy, crying, painful procedure):
    - Increase FiO_2 temporarily
    - Resume supplementary nasal oxygen temporarily if just weaned off oxygen therapy
❑ Top-up transfusions
  To maintain PCV between 40–45 to maximize oxygen delivery to tissues
❑ Ductus arteriosus
  Early closure of duftus reduces pulmonary congestion and facilitates weaning: surgical ligation if indomethacin fails
❑ Fluid restriction
  - Fluid overload exacerbates pulmonary interstitial oedema
  - Closely monitor fluid balance
  - Keep fluid maintenance around 150 ml/kg/day
❑ Drugs to facilitate weaning
  - Diuretics
    - Reduces interstitial oedema
    - Regimen: singly or in combination:
      Furosemide 1–1.5 mg/kg BD (P.O. or IV)
      Chlorothiazide 10–20 mg/kg BD (P.O.)
      Hydrochlorothiazide 1–2 mg/kg BD (P.O.)
      Spironolactone 1–1.5 mg/kg BD (P.O.)
  - Theophylline
    - Effect: diuretic, bronchodilator and respiratory stimulant effect
    - Regimen:

Aminophylline (iv): loading 5–6 mg/kg infusion over 20
min maintenance 4.4 mg/kg infusion over 24 hr
Theophylline (oral): 3–9 mg/kg/day (theophylline level:
5–12 mg/ml)

• Steroid
  – Short-term improvement of lung function facilitating
    weaning
  – Regime (Avery et al., 1985):
    Dexamethasone 0.5 mg/kg/day divided Q12H for 3
    days,
    then 0.3 mg/kg/day divided q12h for 3 days,
    then reduce 10% of dose q 3 days till 0.1 mg/kg/day
    (or 0.2 mg/kg/day × 3 days),
    then 0.1 mg/kg/day on alternate day for 1 week and dis-
    continue,
    given when ventilator dependent at age 3–6 weeks.
  – Problems:
    Optimal dose and duration unclear
    Less effective if used late
    Rebound after taking off steroid been reported
    Risk of sepsis especially disseminated CMV, candida,
    bacterial septicaemia
    Other effects: hyperglycaemia, hypertension, possible
    adrenal supression

❑ Bronchospasm
  • Crying, vigorous suctioning, excessive instilled saline can
    trigger bronchospasm spells
  • Nebulized b-sympathomimetics may provide relief:
    Salbutamol 0.5% (5 mg/ml) 0.05–0.15 mg/kg every 2–4 hour
    Terbutaline 0.1% (1 mg/ml) 0.1–0.3 mg/kg every 2–6 hour
    Metaproterenol 5% (50 mg/ml) 0.25–0.5 mg/kg every 2–4
    hour

❑ Intercurrent infection
  • To send tracheal aspirates for culture and gram stain once to
    twice per week while intubated to allow early dx
  • To treat with antibiotics according to sensitivity guide

4.3 Nutritional support
❑ Nutrition is the key to pulmonary recovery
❑ Provide 120–150 cal/kg/day in recovery phase to meet needs for
  continual growth, tissue repair and increased oxygen consumption due to increased respiratory work
❑ Use high caloric special premie formula (80 cal/100 ml) when
  necessary
❑ To step up oral feeding slowly according to tolerance, supple-
  menting with parenteral feeding

❏ Give supplementary oxygen during oral feeding according to SaO_2
❏ Train oral feeding by occupation therapist

4.4 Complications

Problems	Management/Action
Systemic hypertension	— regular BP monitoring
Pulmonary hypertension	— monthly ECG, XRC — echocardiogram
Recurrent apnoea and bradycardia	— prolonged respiratory monitoring
Gastroesophageal reflux	— upright position; thicken feeds; cisapride
Anaemia	— weekly PCV
Rickets	— anticipate presentation at 2–3 months — regular Ca, PO_4, alkaline phosphatase screen — ensure adequate Ca and PO_4 intake
Nephrocalcinosis	— renal ultrasound screening if on chronic diuretic therapy
Inguinal herniae	— watch for strangulation — elective surgical repair before discharge
Growth failure	— chart growth parameters in conjunction with caloric and third intake
Retinopathy of prematurity	— arrange ophthalmological exam by ophthalmologist by 34–36 weeks post-conceptional age (even if still on oxygen)
Hearing impairment	— hearing screen prior to discharge

References

1. Ban Calari et al. J Pediatr 1979; 95:819–23.
2. Northway et al. New Engl J Med 1967; 276:357.
3. Edwards DK. *The Radiology of Bronchopulmonary dysplasia and its complications.* In Merritt TA, Northway WA et al. (eds): Bronchopulmonary Dysplasia. Boston: Blackwell, 1988; pp. 185–234.

Table 4. Chronic Lung Disease in Premies: Classical Syndromes

Features	BPD	Chronic Pulmonary Insufficiency of Prematurity	Wilson–Mikity Syndrome
Patient group	Usually premie, not specific	Premie < 1250 g	< 32 wk, < 1500 g
Antecedent events	IPPV & oxygen therapy for (a) HMD (b) recurrent apnoeas	Initially well with normal chest X-ray	Initially well with normal chest X-ray
Time of onset	Insidiously following on RDS or after IPPV by 2nd to 3rd wk	End of 1st wk to 2nd wk	2nd–3rd wk
Clinical manifestations	Chronic respiratory failure Failure to wean from ventilator	Insidious respiratory distress Apnoeas Atelactasis Hypoxaemia & CO_2 retention	Tachypnoea, retractions, cyanosis, apnoeas, hypoxaemia, CO_2 retention
Radiology	Classical evolution (Northway staging) — not uniformly seen Spectrum of hyperlucency & strands of radiodensity usually clear in 1–2 years	Underaerated lungs with diffuse haziness	*Early* Bubbles amidst diffuse streaky inflatrates *Late* Hyper-inflated lung bases with infiltrates in upper fields. Rib fractures clear in 6–24 months

Table 4. Chronic Lung Disease in Premies: Classical Syndromes (continued)

Features	BPD	Chronic Pulmonary Insufficiency of Prematurity	Wilson–Mikity Syndrome
Natural course	Chronic respiratory difficulty Ventilator & O_2 dependent for months Mortality 20–40%	Usually recover in 2nd month Mortality 10–20%	May progress to respiratory failure in 4–8 wks Slow recovery in 6 months to 2 years mortality 25–50%
Likely cause	Epithelial damage & repair: immaturity + barotruma + oxygen toxicity ± infection	Progressive atelactasis due to increased chest wall compliance	Uneven airtrapping & atelactasis behind easily collapsible & compliant airways
Px. original description	Northway et al. *Oxygen & IMV supportive care essential.* New Eng J Med 1967; 276:357.	Krauss et al. *CPAP, oxygen, supportive care.* Peds 1975; 55:55.	Wilson M, Mikity V. *Oxygen, IPPV, supportive care.* Amer J Dis Child 1960; 99:489.

10

Persistent Foetal Circulation

Paul KL LAM

1. Definition

1.1 a. Persistent foetal circulation: when there is no predisposing factor.
b. Persistent pulmonary hypertension: secondary to parenchymal lung diseases such as meconium aspiration, severe pneumonia.

1.2 Occurs mainly in term and post-term infants in the newborn period.

1.3 Characterized by severe pulmonary hypertension causing right-to-left shunt through the ductus arteriosus and/or the foramen ovale.

2. Diagnosis

2.1 **History:** there may be history of postmaturity, birth asphyxia, and meconium stained amniotic fluid.

2.2 **Onset:** range from immediate at birth to 24 hours of age.

2.3 **Presentation:** present as cyanosis, simulating cyanotic congenital heart; $PaO_2 < 45$ mm Hg in 100% inspired oxygen.

2.4 **Useful diagnostic tests**
❑ Simultaneous preductal and postductal arterial sampling
  • Obtain preductal blood samples from the right brachial, right radial or temporal artery.
  • Obtain postductal samples from the posterior tibial artery or via an umbilical arterial catheter positioned in the descending aorta distal to the ductus arteriosus.
  • If ductal shunting occurs, the preductal PaO_2 is higher than the postductal PaO_2 by > 15–20 mm Hg.
❑ Simultaneous preductal and postductal $TcPO_2$
  • Measure preductal $TcPO_2$ with sensor placed over right upper chest or shoulder.

- Measure postductal TcPO$_2$ with sensor placed over abdomen or lower limb.
- A preductal TcPO$_2$ higher than postductal TcPO$_2$ demonstrates the presence of ductal shunting.

❏ Hyperoxia-hyperventilation test
- This test can be carried out on infants being mechanically ventilated.
- Manually hyperventilate infant with 100% oxygen using an anaesthesia bag for 5 to 10 minutes.
- At a low PaCO$_2$ of less than 20–30 mm Hg, if the PaO$_2$ rises abruptly to > 100 mm Hg, this test distinguishes PFC from cyanotic congenital heart disease which would not show this response.

❏ Echocardiography
- Ratio of right ventricular pre-ejection period to right ventricular ejection time: prolonged in PFC.
- Contrast echocardiography: after rapidly injecting 0.25 to 0.5 ml.kg 5% dextrose in water through the umbilical vein or peripheral vein of a lower limb, the presence of contrast visualized in both the right ventricular outflow tract and the left atrium demonstrates the atrial right to left shunt.

3. Treatment

3.1 Supportive: correct any acidosis, anaemia, polycythaemia, hypotension, hypocalcaemia, and hypoglycaemia.

3.2 Ventilator therapy
- ❏ Hyperventilation: to achieve a "critical PaCO$_2$" of around 20 to 30 mm Hg, at which a drastic improvement of PaO$_2$ may occur.
- ❏ Keep PaO$_2$ > 100 mm Hg, pH > 7.5 by using sufficiently high pressure and high ventilatory rate. Beware of airleaks.

3.3 Pharmacologic therapy
- ❏ Dopamine: a dosage of 5–15 mg/kg/min may be used to elevate systolic blood pressure.
- ❏ Tolazoline
  - As last resort if above measures fail.
  - Loading dose 1 to 2 mg/kg intravenously through a peripheral vein (preferably over the upper limbs) followed by an IV infusion of 1 to 2 mg/kg/hour.
  - Arterial pressure must be continuously monitored. If hypotension occurs, give volume expanders (blood or plasma) and, if necessary, dopamine infusion.
  - Beware of renal failure and gastrointestinal bleeding which are side effects of tolazoline.

Reference

1. Fox WW and Duara S. *Persistent pulmonary hypertension in the neonate. Diagnosis and management.* J Pediatr 1983; 103:505–14.

11

Congenital Heart Diseases

LEUNG Nin-ming

1. Presentations in neonatal period

Cardiogenic shock
Heart failure
Cyanosis
Asymptomatic heart murmur

1.1 Presenting in first week of life with heart failure or shock
- ❏ Hypoplastic left heart syndrome
- ❏ Aortic arch interruption
- ❏ Coarctation of aorta
- ❏ Other complex structural defects
- ❏ Cardiac arrhythmia
- ❏ Myocarditis
- ❏ Myocardial ischaemia
- ❏ Cardiomyopathy
- ❏ Large VSD and PDA*

1.2 Presenting with early cyanosis
- ❏ Transposition of great arteries
- ❏ Pulmonary atresia
- ❏ Tricuspid atresia
- ❏ Tetralogy of Fallot (Severe)
- ❏ Obstructed total anomalous pulmonary venous drainage

Note: Many non-cardiac conditions have similar clinical presentation as congenital heart diseases.

---

* Heart failure uncommon in neonatal period (due to high pulmonary vascular resistance) except in preterms in whom the pulmonary vascular musculature is less well developed.

2. Points to note in physical examination

2.1 Central cyanosis
- ❏ Polycythaemia may cause cyanosis despite normal oxygen tension.
- ❏ Cyanosis on crying in first few hours after birth may be normal.
- ❏ Nitrogen wash out test: obtain blood for PO_2 determination from right radial artery after breathing 100% oxygen for 10 minutes: cyanotic heart disease unlikely if $PaO_2 > 150$ mm Hg.

2.2 Peripheral pulses: palpate for peripheral pulses on all four limbs. Coarctation of aorta is a cause of heart failure in early infancy.

2.3 Liver size: liver edge down to 3 cm below costal margin may be normal in neonates.

2.4 Blood pressure: always measure blood pressure. Hypertension can be a cause of heart failure.

2.5 Heart sounds: second heart sound is loud and single at birth. Splitting can be easily heard by 48 hours of age in 75% of newborns.

2.6 Bruit: auscultate cranium and abdomen routinely for bruit: large arteriovenous fistula can cause heart failure.

2.7 Murmurs: often difficult to interpret in newborns as some murmurs are transient, e.g., that produced by decrease in pulmonary resistance before ductal closure.
- ❏ Soft ejection systolic murmur is often non-pathological (functional murmur).
- ❏ Pansyotolic murmur at lower left sternal border may be produced by transient tricuspid regurgitation resulting from myocardium ischaemia.
- ❏ Stenotic lesions usually produce murmur early; septal defects produce murmur only later when pulmonary pressure drops.

3. Investigation

3.1 Chest X-ray
- ❏ Cardiothoracic ratio: normal in newborn if < 0.6.
- ❏ Pulmonary vascularity: difficult to evaluate in newborns; impossible to determine whether active or passive congestion is present.

3.2 Electrocardiogram: Criteria for ventricular hypertrophy in neonates:
- ❏ Right ventricular hypertrophy:
  - qR in V_1
  - R in $V_1 > 28$ mm
  - R in $V_4R > 19$ mm

- S in V_6 > 10 mm
- T wave in V_1 positive after day 4
❑ Left ventricular hypertrophy:
  - S in V_1 > 21 mm (term babies)
  - > 26 mm (premature babies)
  - R in V_6 > 16 mm
  - Q wave in V_6 > 3 mm
  - QRS axis > +30°

Notes:
(1) Many structural heart defects give rise to little or no cardiac hypertrophy in utero.
(2) Many respiratory conditions are associated with mild right ventricular hypertrophy.
(3) Relative right ventricular predominance with right axis deviation is normal in newborns.
(4) T wave in V_1 may be upright on day 1, usually becomes negative by day 3.
(5) Premature babies: more LV predominance than in full terms.

3.3 Echocardiogram and doppler studies: mainstay of investigations for cardiac defects.

3.4 Cardiac catheterization: has once been the gold standard for cardiac investigations; used much less frequently nowadays.

4. Management

4.1 Heart failure
❑ Fluid restriction: can go down to 50 ml/kg/day.
❑ Diuretics
❑ Digitalization
❑ Other inotropes, e.g., dopamine especially in severe heart failure.
❑ Vasodilators (e.g. nitroprusside, hydralazine, captopril): for afterload reduction.
   Contraindications: conditions with fixed obstruction and fixed cardiac output (e.g. cardiac tamponade, complete heart block).
❑ Assisted ventilation (for severe heart failure) to:
  - Reduce ventilation work of infant.
  - Decrease pulmonary oedema by using added positive end-expiratory pressure.
❑ Oxygen to relieve hypoxia.
❑ High caloric diet by tube-feeding.
❑ Correct any predisposing factors, e.g., infection.
❑ Correct any correctable structural lesions.

4.2 Maintenance of ductal patency with prostaglandin E: required in coarctation of aorta and many ductus-dependent cyanotic heart diseases.

❑ Dosage:
- IV prostaglandin E_1 or E_2 0.003–0.01 µg/kg/min, or
- Oral prostaglandin E_2 20–25 µg/kg/hour; may be decreased after 1–2 weeks gradually to every 4–6 hours.

❑ Side effects:
- Apnoea
- Hypotension
- Cutaneous vasodilatation
- Seizure
- Fever
- Diarrhoea
- Cortical hyperostosis*
- Friability of ductus arteriosus*
- Damage to pulmonary vascular smooth muscle*

(* After long-term treatment)

4.3 Shunting operation to improve pulmonary blood flow

Usually in form of modified Blalock-Taussig Shunt (MBTS) — subclavian artery and pulmonary artery of the same side connected via an artificial tube.

5. Specific lesions

5.1 Patent ductus arteriosus (PDA)

❑ Pathology/pathophysiology:
- Functional closure of ductus occurs normally in the first day of birth in term infants.
- Delayed closure results from conditions that lower arterial PO_2 or increase prostaglandin production, e.g., asphyxia, meconium aspiration, primary pulmonary diseases.
- Ductus that persists after the 3rd month of life is usually due to structural abnormalities. Chance of spontaneous closure after this stage is < 10%.
- Common in premature infants as the ductal tissue is immature.

❑ Management of PDA in prematures:
- Fluid restriction
  Diuretics ± digoxin
  Maintain haematocrit > 45%
- If above fails in 2 days, give 3 doses of IV or oral indomethacin 12 hours apart:
  Age < 48 hours: 0.2 mg/kg/dose as first dose
                                    0.1 mg/kg/dose in second and third dose

Age 2–7 days: 0.2 mg/kg/dose × 3 doses
Age ≥ 8 days: 0.25 mg/kg/dose × 3 doses
Contraindications: renal failure, bleeding tendency, necrotizing enterocolitis, significant neonatal jaundice.
- Surgical closure if indomethacin fails to close the ductus.

5.2 Coarctation of aorta
❑ Pathology/pathophysiology:
- Preductal type — usually associated with decreased blood flow from left ventricle due to aortic stenosis, mitral stenosis, malaligned VSD, or hypoplasia of descending aorta.
- Juxtaductal type may result from encircling of aorta by abnormal ductal tissues.

❑ Presentations: heart failure — a relatively common cause of heart failure in early neonatal period.

❑ Management:
- Mainly surgery — subclavian flap repair ± pulmonary banding (when large VSD coexist).
- Balloon angioplasty has been tried recently.
- Preoperative medical treatment of heart failure with prostaglandin E ± dopamine ± mechanical ventilation.

5.3 Hypoplastic left heart syndrome
❑ Pathology: Hypoplasia or atresia of left ventricle, mitral valve, aortic valve. Ascending aorta is smaller than aortic arch and systemic circulation depends on ductal supply.

❑ Presentation: presents early with cardiogenic shock.

❑ Management: conservative. Result of surgical palliation is very poor.

5.4 Transposition of great arteries (TGA)
❑ Presentations:
- early cyanosis
- heart failure (those with adequate mixing: e.g. when large VSD coexists)

❑ Management:
- Prostaglandin E before septostomy to maintain ductal patency if severe hypoxia or acidosis present.
  Continue after septostomy if oxygenation does not improve.
- Palliative operation: balloon atrial septostomy
  Oxygenation may not improve immediately after septostomy even with satisfactory atrial communication due to delayed fall in pulmonary vascular resistance, unfavourable ventricular compliance relationship, and unfavourable pattern of intercirculatory streaming.
- Definitive operation: atrial or arterial switch operation
  Early arterial/atrial switch operation required if $PaO_2 < 30$ mm Hg after septostomy.

Atrial switch operation should be carried out early (e.g. several months of age) as TGA is associated with early onset pulmonary vasculature disease.

5.5 Pulmonary atresia with intact ventricular septum (PA + IVS)

❑ Pathology: RV is usually small; outlet portion may be present; there may be tricuspid regurgitation or stenosis.
❑ Presentations:
  • Presents usually early with cyanosis.
  • Chest X-ray may show cardiomegaly due to enlarged RA & LV.
  • ECG usually shows left ventricular predominance.
❑ Management:
  • Prostaglandin E: to maintain ductal patency.
  • Operation: Pulmonary valvotomy ± shunt (balloon atrial septostomy if outlet portion present. Shunting operation alone if there is no outlet.

5.6 Pulmonary atresia + V.S.D. (PA + VSD)

❑ Pathology: Pulmonary blood flow is supplied by PDA or major aortopulmonary collaterals.
❑ Presentations:
  • Cyanosis: if there is decreased pulmonary blood flow.
  • Occasionally presents with heart failure due to increased in pulmonary blood flow.
❑ Management:
  • Shunting operation for those with marked hypoxia or when pulmonary blood supply comes from ductus alone.
  • Ligation or banding of collaterals for those with heart failure.
  • Total correction is very difficult if main pulmonary artery is not present or very hypoplastic.

5.7 Tricuspid atresia (TA)

❑ Pathology: Tricuspid valve is usually absent with normally related great vessels. RV is hypoplastic with VSD.
❑ Presentations:
  • Cyanosis especially when there is pulmonary stenosis.
  • Heart failure is common if it is associated with transposition of great vessels. In these cases, there may be associated subaortic stenosis or even coarctation of aorta.
  • ECG — typically left axis deviation with LV hypertrophy.
❑ Management:
  • Shunting operation if there is significant cyanosis. Balloon atrial septostomy is needed if there is obstructive foramen ovale.
  • Pulmonary artery banding for those with heart failure.
  • Fontan operation (connecting RA to pulmonary artery) is the definitive operation for older children.

5.8 Total anomalous pulmonary venous drainage (TAPVD)

❑ Presentations:
- Obstructed TAPVD is an uncommon cause of early cyanosis with respiratory distress (usually infra-diaphragmatic type). Chest X-ray shows pulmonary oedema with a small heart. A high index of suspicion is needed for diagnosis (e.g. in term infants with severe RDS).
- May sometimes be associated with early heart failure.

❑ Management: surgical

5.9 Atrial isomerism: A group of complex heart disease in which both atria are of similar morphology.

❑ Pathology:
- Liver is central in 50% of right atrial isomerism and about 1/3 of left atrial isomerism.
- Asplenia is usually present in right atrial isomerism and polysplenia in left atrial isomerism.
- Interruption of IVC with azygous continuation of IVC is present in a large proportion of left isomerism.
- The bronchial morphology is similar on both sides.
- TAPVD is common in right atrial isomerism. Common AV value is common in atrial isomerism.
- Right atrial isomerism is commonly associated with single ventricle, pulmonary outflow obstruction and transposition of great arteries.

❑ ECG: 80% of patients with left atrial isomerism have a superiorly orientated P wave axis.

5.10 Ventricular septal defect (VSD)

❑ Presentations:
- Usually asymptomatic in early neonatal period.
- Murmur can be heard when pulmonary pressure drops.
- Drop in pulmonary pressure is much delayed in infants with large VSD.
- Heart failure usually occurs in the first 3 months of age. Close observation during this period is important if the defect is large.

5.11 Tetralogy of fallot (TOF)

❑ Presentations:
- Cyanosis uncommon in neonatal period: TOF usually presents with a high pitched systolic ejection murmur during this period.
- Cyanosis in neonatal period suggests the presence of severe hypoplasia or atresia of pulmonary artery.
- Cyanotic attacks are rare in the first month of life.

❑ Management: Prostaglandin E infusion and shunting operation for those who present with early cyanosis.

References

1. Anderson, Macartney, Shinebourne and Tynan (eds). *Paediatric Cardiology*. 1st ed. London, Churchill Livingstone, 1987.
2. Adams, Emmanouilides and Riemenschneider (eds). *Moss' Heart Disease in Infants, Children and Adolescents*. 4th ed. Williams & Wilkins, 1989.

12
Asphyxia Neonatorum

Betty WY YOUNG

1. Definition

- ❏ The end result of reduced oxygen and nutrient supply to the foetal brain.
- ❏ Little agreement on the clinical definitions which include:
  - Delay in establishing spontaneous respiration
  - Low Apgar score
  - Cord blood acidosis
  - Cardiotocograph abnormalities
  - Disturbance in behaviour of infant following birth

2. Neuropathological processes associated with perinatal asphyxia

- ❏ Primary subarachnoid haemorrhage
- ❏ Hypoxic-ischaemic encephalopathy (HIE)
- ❏ Periventricular leukomalacia
- ❏ Periventricular, intraventricular haemorrhage

3. Hypoxic-ischaemic encephalopathy (HIE)

It is the most reliable indicator of asphyxia and major determinant of neurologic morbidity and mortality in neonatal period and later in childhood.

Severity of HIE

Grade I (mild)	Grade II (moderate)	Grade III (severe)
"Hyperalertness"	Lethargy	Coma
Mild hypotonia	Marked abnormalities in tone	Severe hypotonia
Poor sucking	Require tube feeds	Failure to maintain spontaneous respiration
Uninhibited reflexes	Suppressed primitive reflexes	Suppressed brain-stem function
No seizures	Seizures	Prolonged seizures
Sympathetic over-activity — Mydriasis — Tachycardia	Parasympathetic over-activity — Pinpoint pupil — Relative bradycardia — Copious secretion	
Duration < 24 hours	Signs of recovery at end of 1st week	

4. Complications

4.1 **Cardiovascular:** Cardiac failure, hypotension, myocardial ischaemia

4.2 **Renal:** Acute tubular necrosis, oliguria

4.3 **Pulmonary:** Meconium aspiration, respiratory distress secondary to cardiac failure, persistent pulmonary hypertension

4.4 **Metabolic:** Hyponatraemia, hypoglycaemia, hypocalcaemia, metabolic acidosis, hyperammonaemia

4.5 **Haematological:** Disseminated intravascular coagulation

4.6 **Gastrointestinal:** Stress ulcers, necrotizing enterocolitis

5. Management

5.1 **Prevention of intrauterine asphyxia**
   ❏ Antepartum foetal assessment:
   • Foetal body movements
   • Foetal pupillary responses and eye movements (Ultrasound)
   • Foetal heart rate — Non-stress test
                      — Contraction stress test
   • Amniotic fluid volume (Ultrasound)
   • Foetoplacental blood flow (Doppler)

❑ Intrapartum foetal assessment:
- Early passage of "heavy" meconium in utero
- Electronic foetal heart rate monitoring
- Foetal acid-base status

5.2 General support

❑ Respiratory:
- Mechanical ventilation if:
  - Severe respiratory distress complicating meconium aspiration
  - Coma after severe asphyxia
  - Severe and prolonged convulsion
  - Hypoxia and hypercapnia

❑ Circulatory:
- Correct systemic hypotension:
  Colloid 10–20 ml/kg/IV
  If hypotension is secondary to myocardial ischaemia:
  Dopamine 5 µg/kg/min increasing to 10, 15 or 20 µg/kg/min if mean arterial pressure < 40 mm Hg
- Treat hyperviscosity secondary to polycythaemia by partial exchange transfusion with fresh frozen plasma or 5% albumin.

  Volume of exchange (ml) =

  $$\text{Blood volume} \times \frac{\text{Observed haematocrit} - \text{Desired haematocrit}}{\text{Observed haematocrit}}$$

❑ Haemostasis:
- If coagulopathy present:
  - Supportive management, no place for systemic heparinization
  - Additional Vit K_1
  - Transfusion of fresh frozen plasma

❑ Metabolic:
- Hypoglycaemia:
  - Maintain blood glucose level 75–100 mg/dl
  - Avoid hyperglycaemia and rapid bolus injection of concentrated glucose
- Acidosis: Slow infusion of 4.2% $NaHCO_3$

❑ Fluid:
- Fluid restriction:
  - Prevent brain oedema
  - Avoid fluid overload if oliguria present
- Maintain serum osmolality around 290 mOsm/l and urinary SG at 1010

5.3 Cerebral treatment
❏ Anticonvulsant drugs

	Loading dose	Maintenance dose	Therapeutic blood level (μg/ml)
Phenobarbitone	20 mg/kg IV	6 mg/kg 12 hourly dose	20–40
Phenytoin	20 mg/kg IV	5–8 mg/kg 12 hourly dose	10–20
Paraldehyde	0.1–0.2 ml/kgim	—	—
Diazapam	0.3 mg/kg IV	—	—
Clonazepam	0.25 mg	0.1 mg 12 hourly doses	Not helpful

❏ Intracranial hypertension
  • Hyperventilation:
    – Hypocapnoea will reduce cerebral blood flow which seems undesirable
    – Maintain $PaCO_2$ 3.5–4.0 KPa
  • Corticosteroids: Controversial, not recommended
  • Osmotic agents:
    – 20% Mannitol 1 gm/kg over 20 minutes
    – Theoretical hazard of entry into brain causing rebound effect of brain swelling
  • Barbiturates:
    – Little protective effect when given after asphyxia
    – Thiopental: side effect of dramatic fall in blood pressure
  • Elevation of head

6. Bad prognostic factors

6.1 Abnormal foetal heart rate in the two hours before delivery, especially prolonged foetal bradycardia
6.2 Base deficit > 10 mEq/l in first 60 minutes of life
6.3 Depressed Apgar score
  ❏ Unreliable as the score may be affected by:
    • Interobserver variability
    • Effect of drugs given to mother before delivey
    • Reversible stress of delivery
  ❏ Apgar score < 4 at 20 minutes, 57% chance of cerebral palsy in survivors
6.4 Delayed onset of respiration by 30 minutes
  ❏ Affected by other factors

- Effect of drugs given to mother
- Neuromuscular disorder of newborn

6.5 Neonatal neurologic syndrome
❏ Grade III encephalopathy
❏ Abnormal neurological features persisting beyond 1–2 weeks
❏ Seizures: Early onset and difficult to control
❏ Increased intracranial pressure > 15mm Hg and resistant to treatment
❏ Persistent abnormality of brain stem function

6.6 EEG assessment
❏ EEG abnormalities
  - Isoelectric recordings
  - Periodic patterns
  - Persistent low voltage states
❏ Abnormal auditory evoked brain stem responses

6.7 CT scan
❏ Extensive hypodensity in white and grey matter
❏ Intracranial haemorrhage

References

1. Levene M. *Neonatal Neurology. Current Reviews in Paediatris.* Churchill Livingstone, 1987.

2. Fenichel GM. *Neonatal Neurology.* 2nd ed. Clinical Neurology & Neurosurgery Monographs. Churchill Livingstone, 1985.

3. Volpe JJ. *Neurology of the Newborn.* 2nd ed. WB Saunders, 1987.

4. Fanaroff AA & Martin RJ. *Neonatal-Perinatal Medicine. Diseases of the Fetus & Infant.* 4th ed. CV Mosby Co, 1987.

5. Hill A & Volpe JJ. *Perinatal asphyxia: Clinical aspects.* Clin Perinatol 1989; 16(2):435–57.

6. Hill A. *Assessment of the fetus: Relevance to brain injury.* Clin Perinatol 1989; 16(2):413–34.

7. Perlman JM. *Systemic abnormalities in term infants following perinatal asphyxia. Relevance to long term neurologic outcome.* Clin Perinatol 1989; 16(2):475–84.

8. Mizrahi EM. *Consensus & controversy in the clinical management of neonatal seizures.* Clin Perinatol 1989; 16(2):485–500.

9. Jacob MM & Phibbs RH. *Prevention, recognition & treatment of perinatal asphyxia.* Clin Perinatol 1989; 16(4):785–807.

10. Bedrick AD. *Perinatal asphyxia and cerebral palsy. Fact, fiction or legal prediction?* Editorial AJDC 1989; 143:1139–40.

11. Hall DMB. *Birth asphyxia and cerebral palsy.* BMJ 1989; 143:279–82.

12. Vannucci RC. *Current and potentially new management strategies*

for perinatal hypoxic — ischaemic encephalopathy. Paediatrics 1990; 85(6):961–68.

13. Portman RJ, Carter BS, Gaylord MS et al. *Predicting neonatal morbidity after perinatal asphyxia: A scoring system.* Am J Obstet Gynecol 1990; 162:174–82.

14. Gilstrap LC, Leveno KJ, Burris J et al. *Diagnosis of birth asphyxia on the basis of fetal pH, Apgar score and newborn cerebral dysfunction.* Am J Obstet Gynecol 1989; 161:825–30.

15. Whitelaw A. *Intervention after birth asphyxia.* Arch Dis Child 1989; 64:66–68.

13

Periventricular Haemorrhage (PVH)

FOK Tai-fai

1. Incidence

35–50% in infants < 1500 g
Up to 65% in infants < 1000 g

2. Associating factors

2.1 Extreme prematurity and low birth weight are the major associating factors.

2.2 Other possible associating factors include:
- ❏ Hypoxia
- ❏ Hypercapnia
- ❏ Birth asphyxia
- ❏ Acidosis
- ❏ Respiratory distress syndrome
- ❏ Assisted ventilation
- ❏ Fluctuation in arterial blood pressure
- ❏ Infusion of $NaHCO_3$

3. Clinical features

3.1 Mostly occur within first week (first 3 days) of life.

3.2 Minor haemorrhages mostly asymptomatic.

3.3 Subtle neurological signs such as apnoea, impaired motility and hypotonia.

3.4 Massive bleeding may result in bulging anterior fontanelle, seizure, coma, respiratory arrest, and death.

4. Diagnosis

Done by cranial ultrasound which should be carried out regularly and whenever the baby deteriorates.

4.1   Grading: Several grading systems have been proposed. Examples include:

❏ Papille et al (1978):
  • Grade 1: Isolated subependymal haemorrhage (SEH)
  • Grade 2: Rupture into lateral ventricle without ventricular dilatation
  • Grade 3: Rupture into lateral ventricle with ventricular dilatation
  • Grade 4: Intraventricular haemorrhage (IVH) with parenchymal extension

❏ Lazzara et al (1980):
  • Mild: SEH ± ventricular blood filling ≤ 25% of the ventricles
  • Moderate: IVH with blood filling 25–50% of the ventricles
  • Severe: IVH with blood filling ≥ 50% of the ventricles

❏ Levene et al (1982):
  • Grade 1: Haemorrhage in region of germinal matrix with no extension
  • Grade 2: Downward extension or thrombus in lateral ventricle
  • Grade 3: Parenchymal extension

❏ Shankaran et al (1982):
  • Mild: SEH ± small amount of blood in normal sized ventricles
  • Moderate: Intermediate amount of blood in enlarged ventricles
  • Severe: IVH filling entire lateral ventricle forming a cast ± parenchymal haemorrhage

❏ Levene et al (1983): PVH and ventricular dilatation are graded separately.
  • Haemorrhage:
    0 — No haemorrhage
    1 — Localized haemorrhage ≤ 1 cm in its largest diameter
    2 — Haemorrhage > 1 cm in its largest measurement but not extending beyond the atrium of the lateral ventricle
    3 — Blood clot forming a cast of the lateral ventricle and extending beyond atrium
    4 — Intraparenchymal haemorrhage
  • Ventricular dilatation:
    0 — No dilatation
    1 — Transient dilatation
    2 — Persistent but stable dilatation
    3 — Progressive dilatation requiring treatment
    4 — Persistent asymmetrical ventricular dilatation

❏ Hence when grading PVH, the grading method needs to be specified. One simple way for grading PVH and ventricular dilatation is:

- Haemorrhage:
  Grade 1: Subependymal haemorrhage
  Grade 2: IVH filling < 50% of lateral ventricle
  Grade 3: IVH filling > 50% of lateral ventricle
  Grade 4: IVH with parenchymal extension
- Ventricular dilatation:
  Grade 0: No dilatation
  Grade 1: Transient dilatation
  Grade 2: Persistent but stable dilatation
  Grade 3: Progressive dilatation requiring treatment

❏ Ventricular dilatation may be diagnosed ultrsonographically by measuring the ventricular index (VI — distance from the falx to the lateral-most point of either lateral ventricle measured on coronal scan at the level of hippocampal echo just posterior to the foramen of Monro).

Ventricular dilatation is considered to be present when VI exceeds the 97th centile on the centile chart for VI (Levene MI, 1981).

References

1. Papille LA, Burstein J, Burstein R, et al. *Incidence and evolution of subependymal and intraventricular hemorrhage: a study of infants with birth weight less than 1500 gm.* J Pediatr 1978; 92:529–34.

2. Lazzara A, Ahmann P, Dykes F, et al. *Clinical predictability of intraventricular hemorrhage in preterm infants.* Pediatrics 1980; 65:30–34.

3. Levene MI, Fawer C-L, Lamont RF. *Risk factors in the development of intraventricular haemorrhage in the preterm neonate.* Arch Dis Child 1982; 57:410–17.

4. Shankaran S, Slovis TL, Bedard MP, et al. *Sonographic classification of intracranial hemorrhage. A prognostic indicator of mortality, morbidity, and short-term neurologic outcome.* J Pediatr 1982; 100:469–75.

5. Levene MI, de Crespigny LCh. *Classification of intraventricular haemorrhage.* Lancet 1983; 1:643.

6. Levene MI. *Measurement of the growth of the lateral ventricles in preterm infants with real-time ultrasound.* Arch Dis Child 1981; 56:900–904.

14

Post-haemorrhagic Ventricular Dilatation

Fok Tai-fai

1. Definition

1.1 Hydrocephalus: a descriptive term to describe a clinical condition with features including rapid head growth, dilated scalp veins, sunsetting eyes with or without other features of increased intracranial pressure such as lethargy and vomiting.

1.2 Ventriculomegaly: Ventricular dilatation not progressing to hydrocephalus.

2. Management of post-haemorrhagic ventricular dilatation

2.1 Indications for treatment

Progressive dilatation of the cerebral ventricular system especially in the presence of:

- ❏ Progressive increase in head circumference in excess of the norm (upward "crossing" of the percentile lines), and/or
- ❏ Evidence of increased ICP:
  - Full or tense anterior fontanelle,
  - Separation of the sutures,
  - Clinical manifestations such as seizure, apnoea, bradycardia.

2.2 Medical treatment

May be tried in slowly progressive hydrocephalus.

Should not be relied on when there is evidence of increased ICP.

- ❏ Acetazolamide
  - Dosage: 25 mg/d × 1 day (in 3 divided daily doses), then

increase by 25 mg/d each day up to a maximum of 100 mg/kg/d.
- May be given in combination with frusemide 1–3 mg/kg/d in 3 divided daily doses.
- Complications:
  Acetazolamide: Hyperchloraemic acidosis requiring prophylactic $NaHCO_3$ therapy.
  Frusemide:
    – Metabolic alkalosis (the combination in the doses used will result in a net metabolic acidosis)
    – Hypokalaemia
❑ Isosorbide
- Dosage: 2 g/kg/dose 6 hourly
- Complications: Hypernatraemia, dehydration, diarrhoea, vomiting — interrupt treatment if these occur, resume after recovery with reduced dosage.
- The volume of the drug (1 g dissolved in 2.2 ml liquid) may be prohibitory.

2.3 Serial lumbar puncture
❑ Remove 10–20 ml CSF daily until cessation of progression of ventricular dilatation as documented by cranial ultrasound, then as necessary thereafter.
❑ Useful only when communication between lateral ventricles and subarachnoid space can be confirmed by one of the followings after LP removal of CSF:
- Decrease in ICP at both the anterior fontanelle and LP site, and
- Decrease in ventricular size on cranial ultrasound.

2.4 Serial ventricular tap
❑ Use a lumbar puncture needle with stylet.
❑ Desirable to measure depth of lateral ventricles from scalp with ultrasound first.
❑ Insert needle through the anterior fontanelle at a position just above the eye.
❑ Remove stylet when the predetermined length has been inserted.
❑ Remove 10–15 ml of CSF by free drainage, do not aspirate with a syringe.
❑ Assess adequacy of drainage clinically and by cranial ultrasound. Adequate if:
- Increase in head circumference < 1 cm/week
- Anterior fontanelle normal tension, cranial sutures normal position
- Unchanged or decreasing ventricular size on ultrasound.
❑ Complications:
- Infection

- Damage of brain parenchyma and cerebral cavitation formation
- Intracranial haemorrhage, the risk might be higher if CSF rapidly aspirated
- Hyponatraemia
- Hypoproteinaemia

2.5 Ventricular catheter/reservoir:

❏ A ventricular catheter with a subcutaneous reservoir (e.g. Holter-Hausner catheter, Leroy catheter) is inserted by the neurosurgeon.

❏ Daily removal of 10–15 ml CSF by tapping the reservoir with a 25 or 27 gauge scalp vein needle under aseptic conditions. May repeat 8–12 hourly if anterior fontanelle becomes full or tense.

❏ CSF must be removed very slowly.

❏ Assess adequacy of drainage as in 2.4 and also by intracranial pressure measured through the catheter.

❏ Complications: same as 2.4.

2.6 Ventriculo-peritoneal/ventriculo-atrial shunt

❏ Considered when:
- There is progressive dilatation of the ventricles
- Body weight > 2000 g
- Protein concentration of CSF < 1.5 g/L.

❏ Complications: In general, the younger and smaller the infants, the higher the complication rate and the worse the outcome.
- Wound dehiscence and skin breakdown.
- Infection of the shunt: most commonly by staphylococcus epidermidis.
  When suspected:
  – Blood culture and culture of CSF obtained from the reservoir of the shunt.
  – Start IV broad spectrum antibiotics (consider vancomycin), pending culture results.
  – Remove shunt immediately after infection confirmed.
  – Continue antibiotics for at least 7–10 days.
  – Repeat CSF culture 48 hours after stopping antibiotics.
  – Insert a new shunt if CSF culture negative.
- Blocked shunt: more common if CSF protein concentration is high.

References

1.   Shinnar S, Gammon K, Bergman EW, et al. *Management of hydrocephalus in infancy: Use of acetazolamide and frusemide to avoid cerebrospinal fluid shunts*. J Pediatr 1985; 107:31–37.

2. Lorber J. *Isosorbide in treatment of infantile hydrocephalus.* Arch Dis Child 1975; 50:431–36.

3. Kreusser KL, Tarby TJ, Kovnar E, et al. *Serial lumbar punctures for at least temporary amelioration of neonatal posthemorrhagic hydrocephalus.* Pediatrics 1985; 75:719–23.

4. Leonhardt A, Steiner H-H, Linderkamp O. *Management of posthaemorrhagic hydrocephalus with a subcutaneous ventricular catheter reservoir in premature infants.* Arch Dis Child 1989; 64:24–28.

5. James HE, Bejar R, Gluck L, et al. *Ventriculoperitoneal shunts in high risk newborns weighing under 2000 grams: a clinical report.* Neurosurgery 1984; 15:198–202.

15
Seizures

Alex KH CHAN

1. Clinical manifestations

1.1 Multifocal clonic seizures
1.2 Focal clonic seizures
1.3 Tonic seizures
1.4 Myoclonic seizures
1.5 Subtle seizures, which may manifest as
  ❏ Apnoea
  ❏ Ocular signs: staring, eye deviation, nystagmus, blinking
  ❏ Swimming, pedalling, stepping or rotary arm movements
  ❏ Oral-buccal-lingual movements

2. Causes

2.1 Related to delivery:
  ❏ Hypoxic-ischaemic encephalopathy
  ❏ Birth trauma
  ❏ Intracranial haemorrhage
2.2 Infections: Bacterial and viral infections
2.3 Metabolic causes:
  ❏ Hypoglycaemia
  ❏ Hypocalcaemia
  ❏ Hypomagnesaemia
  ❏ Hypo/hypernatraemia
  ❏ Hyperbilirubinaemia
  ❏ Amino acid abnormalities
  ❏ Pyridoxine deficiency
2.4 Drug withdrawal:
  e.g. Narcotics

2.5 Developmental abnormalities of the brain

3. Investigations for neonatal seizures

3.1 Blood tests:
- ❏ Glucose (includes urgent "dextrostix")
- ❏ Electrolytes: calcium, sodium, potassium, magnesium
- ❏ Blood gas
- ❏ Urea and creatinine
- ❏ Complete blood count

3.2 Cerebrospinal fluid for presence of blood, cell count, Gram-stain and bacterial and viral culture

3.3 Blood culture

3.4 Imaging:
- ❏ Cranial Ultrasound
- ❏ CT scan of the brain when ultrasound cannot enable satisfactory visualization of intracranial structures
- ❏ XR chest may be necessary

3.5 Electroencephalography

3.6 Urine screen for amino acids and other metabolic disorders

4. Management of neonatal seizures

4.1 Clear airway, give oxygen if baby breathing but cyanotic, artificial ventilation if baby apnoeic.

4.2 Anticonvulsants:
- ❏ Phenobarbital:
  Loading dose: 10–20 mg/kg IVI
  Maintenance: 1st 2 weeks: 2–4 mg/kg/day IMI/oral
  Subsequently: 5 mg/kg/day
  Therapeutic drug level: 20–40 mcg/ml
- ❏ Diphenyhydantoin:
  Loading dose: 10–20 mg/kg IVI
  Maintenance: 4–6 mg/kg/day IVI/Oral
  Therapeutic drug level: 10–20 mcg/ml
  Note: Slow absorption when given orally — monitor drug level especially when large doses given.
- ❏ Other anticonvulsants:
  - • Primidone:
    Loading dose: 15–25 mg/kg oral
    Maintenance: 12–20 mg/kg/day oral
    Note: used as an adjunct to first line anticonvulsant

- Paraldehyde:
  0.15 ml/kg IMI/rectal 4 hourly prn
  4% solution IV infusion (titrate drip rate)
  Note: may cause pulmonary oedema, haemorrhage and hypotension
- Diazepam: 0.3 mg/kg/hour IVI
  Note: poor drug for maintenance, uncouples bilirubin albumin complex and increase risk of kernicterus
- Pyridoxine: 100 mg IVI
  Note: given with EEG monitoring for diagnosis of pyridoxine deficiency
❏ Duration of anticonvulsant the therapy:
- Seizures due to transient metabolic disturbance: anticonvulsants may be discontinued when the acute illness is over.
- Seizures due to birth asphyxia: consider discontinuing anticonvulsant after patient has been seizure-free for 2 weeks.
- Indications for longer period of treatment:
  - Persistent neurological abnormality
  - Persistent EEG abnormality
  - Underlying abnormality of the brain

4.3 Treatment of underlying cause(s) (see chapters on asphyxia neonatorum, electrolyte disturbance, metabolic disorders, neonatal jaundice and sepsis)

References

1. Volpe JJ. *Neonatal seizures: Current concepts and revised classification*. Pediatrics 1989; 84(3):422–28.
2. Mizrahi EM. *Consensus and controversy in the clinical management of neonatal seizure*. Clin Perinatol 1989; 16(2):485–500.
3. *Neonatal seizures*. Lancet 1989; 2:135–37.
4. Painter MJ, Bergman I, Crumrine P. *Neonatal seizures*. Paediatric Clin North Am 1986; 33(1):91–109.
5. Powell C, Painter MJ, Pippenger CE. *Primidone therapy in refractory neonatal seizures*. J Pediatr 1984; 105:651–53.
6. Roberton NRC. *Manual of Neonatal Intensive Care*. 2nd ed. London, Edward Arnold Ltd, 1986; pp. 235–38.
7. Avery GB. *Neonatology Pathophysiology and Management of the Newborn*. 2nd ed. Philadelphia, JB Lippincott, 1981.

16

Hypotonia

1. Neurological maturation

Neurological status is different for babies of different gestational maturity. See Table 1.

2. Assessment of tone and strength

2.1 Observe infant's posture in supine position.

Normal resting posture. Note adduction of the thighs and an attitude of flexion in the limb joints.

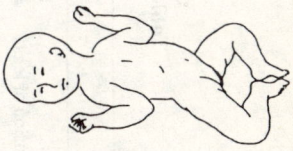

Abnormal resting posture. The thighs are fully abducted (frog-leg), and the arms lie in a flaccid position beside the head.

2.2 Assess resistance to passive movement.
2.3 "Pull to sit" manoeuvre: assessing the head lag in response to traction on the wrists.

Table 1. Neurological Maturation

Function	26 weeks	30 weeks	34 weeks	38 weeks
Resting posture	Flexion of arms Flexion or extension of legs	Flexion of arms Flexion or extension of legs	Flexion of all limbs	Flexion of all limbs
Arousal	Unable to maintain	Maintain briefly	Remain awake	Remain awake
Rooting	Absent	Long latency	Present	Present
Sucking	Absent	Long latency	Weak	Vigorous
Pupillary reflex	Absent	Variable	Present	Present
Traction	No response	No response	Head lag	Mild head lag
Moro	No response	Extension, no adduction	Adduction variable	Complete
Withdrawal	Absent	Withdrawal only	Crossed extension	Crossed extension

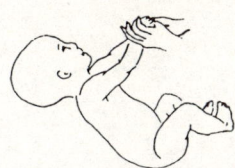

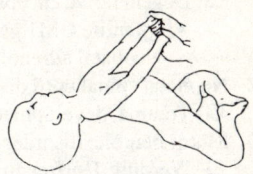

Normal traction response. The lift of the head is almost parallel to the lift of the body and there is flexion in all limb joints.

Abnormal traction response. The head falls backward as the body is pulled forward, and there is no resistance to traction in the arms.

2.4   Observation in ventral suspension

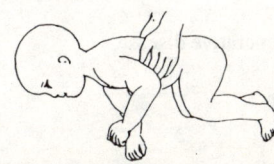

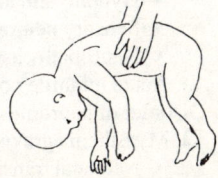

Normal horizontal suspension. There is intermittent raising of the head and flexion of the limbs against gravity.

Abnormal horizontal suspension. The head, body, and limbs hang limply with little or no resistance to gravity.

2.5   Arm and leg recoil

3.   Differential diagnosis of infantile hypotonia

3.1   Cerebral hypotonia
- ❑ Congenital encephalopathy
  - Hypoxia ischaemia
  - Intracranial infection
  - Intracranial haemorrhage
  - Metabolic:
    - Electrolyte, glucose, calcium, magnesium, bilirubin, amino acid, organic acid
    - Drug intoxication
  - Endocrine: Hypothyroidism
  - Developmental disturbance of cerebrum:
    - Cerebral malformation
    - Prader Willi syndrome
    - Cerebrohepatorenal syndrome (Zellweger)
    - Oculocerebrorenal syndrome (Lowe)
    - Other chromosomal disorders

❑ Degenerative encephalopathy
  • Infantile GM1 gangliosidosis
  • Neonatal adrenoleukodystrophy

3.2   Neonatal spinal cord disorder
  ❑ Trauma, developmental

3.3   Motor neurone disorder
  ❑ Werdnig-Hoffmann disease
  ❑ Neurogenic arthrogryposis multiplex congenita
  ❑ Glycogen storage disease type II (Pompe's)
  ❑ Neonatal poliomyelitis (other enterovirus?)

3.4   Peripheral neuropathy-exceedingly rare
  ❑ Chronic polyneuropathy
    • Axonal polyneuropathy
    • Hypomyelinative neuropathy
    • Sensory neuropathy
    • Neuropathy associated with degenerative disease
  ❑ Acute infantile polyneuropathy

3.5   Disorder of neuromuscular transmission
  ❑ Myasthenia gravis
    • Neonatal transient
    • Congenital myasthenia syndrome
    • Familial infantile myasthenia
  ❑ Toxic metabolic
    • Hypermagnesaemia
    • Antibiotic: aminoglycoside
  ❑ Infantile botulism

3.6   Muscle disorder
  ❑ Histology not diagnostic:
    • Congenital myotonic dystrophy
    • Congenital muscular dystrophy
      – Without cerebral involvement
      – With cerebral involvement
    • Polymyositis
    • Minimal change myopathy
  ❑ Histology diagnostic:
    • Central core disease
    • Nemaline (rod body) myopathy
    • Myotubular myopathy
    • Congenital fibre type disproportion
    • Mitochondrial myopathy: cytochrome C oxidase deficiency
    • Metabolic myopathy:
      – Glycogen disorder
      – Lipid disorder: carnitine deficiency

4. Clinical presentations/physical findings

4.1 Cerebral hypotonia
- ❏ Hypotonia more severe than muscle weakness
- ❏ Abnormalities of other brain function
- ❏ Dysmorphic features, malformation of other organs
- ❏ Fisting of hands, scissoring on vertical suspension
- ❏ Normal or brisk tendon reflexes
- ❏ Good recoil strength of limbs on stimulation
- ❏ Improvement of tone with time

4.2 Neuromuscular disease of antenatal onset
- ❏ Diminished fetal movements
- ❏ Polyhydramnios
- ❏ Breech presentation
- ❏ Birth asphyxia
- ❏ Arthrogryposis and associated congenital anomalies
- ❏ Congenital hip dislocation
- ❏ Undescended testis
- ❏ Thin ribs on CXR
- ❏ Hypotonia and muscle weakness
- ❏ Absent or decreased tendone reflexes
- ❏ Fasciculation in motor neurone disease
- ❏ Abnormality of cranial nerves
- ❏ Other congenital anomalies (high arched palate, micrognathia)
- ❏ Parental consanguinity
- ❏ Any family history of neonatal/infant death or neuromuscular disorder
- ❏ Any abnormal neurologic examination of parents, such as myotonia, muscle fatigue

Note: Neonatal neuromuscular disease is often associated with hypoxic ischaemic encephalopathy (HIE) due to failure to establish respiration. Hypotonia in these infants may be attributed to HIE and the underlying neuromuscular disorder is masked.

5. Specific clinical features

5.1 Motor neurone disease (see Table 2).

5.2 Neonatal transient myasthenia gravis
- ❏ Feeding problem
- ❏ Respiratory disturbance
- ❏ Weak cry, facial weakness
- ❏ Generalized muscle weakness, hypotonia
- ❏ Ptosis or oculomotor disturbance

Table 2. Motor Neurone Diseases

	Werdnig-Hoffman disease	Neurogenic arthrogryposis multiplex	Type II glycogen storage disease (Pompe's disease)	Neonatal poliomyelitis
Onset	In utero/birth to several weeks	In utero	Often delayed for weeks	Birth to several weeks
Motor abnormalities	Generalized hypotonia, proximal > distal; muscle atrophy	Generalized or localized (i.e. caudal or cervical)	Prominent, weak muscles	Asymmetric flaccid paralysis
Contractures	Uncommon	Multiple, severe; onset in utero	Uncommon	Rapid postnatal progression
Cranial nerve abnormalities	Disturbed sucking, swallowing, tongue atrophy, fasciculations	Bulbar muscles normal	Disturbed sucking, swallowing; enlarged tongue with fasciculations	May be present
Tendon reflexes	Absent	Absent	Variable	Asymmetric
Clinical course	Rapidly progressive	Nonprogressive	Rapidly progressive	Rapidly progressive
Other organ involvement	None	None	Cardiac, liver, brain	Brain (encephalitis)
Pathogenesis	Autosomal recessive	Sporadic, intrauterine spinal cord dysgenesis	Glycogen deposition in tissues, acid maltase deficiency	Prenatal or postnatal infection by poliovirus

Table 3. Distinguishing features of congenital myasthenic syndromes and neonatal transient myasthenia gravis

	Neonatal transient	"Congenital Myasthenia"	"Familial infantile Myasthenia"
Onset in neonatal period	+	+	+
Myasthenia mother	+	−	−
Familial occurrence	−	+	+
Prominent ophthalmoplegia, ptosis	−	+	−
Generalized weakness	+	−	+
Response to anticholinesterase drugs	+	+	+
Course	Transient, variable	Persistent, mild	Remission of improvement, recurrences

5.3 Congenital myotonic dystrophy
  ❑ Respiratory problem, feeding problem
  ❑ Facial diplegia
  ❑ Hypotonia, areflexia, atrophy
  ❑ Arthrogryposis
  ❑ Transmission via mother
  ❑ Autosomal dominant
5.4 Congenital muscular dystrophy
  ❑ Autosomal recessive
  ❑ Facial weakness
  ❑ Axial weakness, neck and trunk
  ❑ Limb weakness, proximal > distal
  ❑ Contractures
  ❑ Nonprogressive or slowly progressive
  ❑ Some cases with CNS involvement
  ❑ Some cases with swallowing and respiratory problem
5.5 Specific congenital myopathies
  ❑ Autosomal dominant > recessive > X-linked recessive
  ❑ Onset in neonatal period
  ❑ Recognition after the neonatal period
  ❑ Weakness, proximal > distal
  ❑ Tendon reflexes decreased in proportion to weakness

**Table 4. Distinguishing clinical features of
specific congenital myopathies**

	Neonatal hypotonia and weakness	Severe form with neonatal death	Facial weakness	Ptosis	Extra-ocular muscular weakness
Central core disease	+	0	0	0	0
Nemaline myopathy	+	+	+	0	0
Myotubular myopathy	+		+	+	+
Congenital fiber type disproportion	+	0	+	0	0

6. Major laboratory investigation

6.1 Serum creatine phosphokinase (CK)
❑ High level must be interpreted with caution — 10 fold rise may occur in normal infants up to one week following vaginal delivery.
❑ May be increased in acute hypoxic ischaemic encephalopathy, in presence of acidosis.
❑ Not elevated in disorders of the anterior horn cell, peripheral nerve, neuromuscular junction and all muscle disorder.

6.2 Aspartate amino transferase
❑ Elevation may suggest a neuromuscular disease with prominent hepatic involvement.

6.3 Nerve conduction velocity/electromyography (Table 5)
❑ Technical difficulty, interpretation may be confusing.
❑ Myotonic discharges not present.
❑ Repetitive nerve stimulation test not reliable.

6.4 Chest radiograph: look for thin rib.

6.5 Muscle biopsy (Table 6)
❑ For definitive diagnosis to determine prognosis.
❑ For institution of specific and supportive therapy, early biopsy is warranted despite occasional problems in interpretation.

6.6 Sural nerve biopsy
❑ Rarely required except in cases of polyneuropathy.

6.7 Cerebrospinal fluid
❑ CSF protein only raised in peripheral nerve disorder and in some degenerative diseases.

Table 5. Nerve Conduction Velocity/Electromyography Findings in Neuromuscular Disease

Investigation	Observed activity	Site of lesion		
		Anterior horn cell	Peripheral nerve	Muscle
Nerve conduction velocity		N —→	(Demyelination) (Axonal)	N
EMG — muscle at rest	Fibrillation Fasciculation	+	+ ±	— —
EMG — contracted muscle	Amplitude	← →	N —→ N —→ N —→	→ → N —→
	Duration number polyphasics	Increase, larger size	Increased, variable size	Increased, small size

Table 6. Abnormalities in Muscle Biopsies of Patients With Neuromuscular Disease

Abnormality on biopsy	Location of abnormality				
	Central	Anterior horn cell	Peripheral nerve	Neuromuscular junction	Muscle
Fibre atrophy	–	Fiber grouping	Fiber grouping, less severe	–	No grouping
Degeneration of fibres	–	–	–	–	+
Fiber size variation	–	–	–	–	+
Location of nuclei	Peripheral	Peripheral	Peripheral	Peripheral	Central
Increased fat and connective tissue	–	–	–	–	+
Vacuoles					
Glycogen (periodic acid-schiff)	–	–	–	–	+
Lipid (oil red 0 stain)					
Abnormal mitochondria "Ragged-red fibres"	±	–	±	–	+

References

1. Fenichel GM. *Neonatal Neurology.* 3rd ed. New York, Churchill Livingstone, 1990.
2. Roland EH. *Neuromuscular disorders in the newborn.* Clin Perinatol 1989; 16:519–48.
3. Volpe JJ. *Neurology of the Newborn.* 2nd ed. Philadelphia, WB Saunders, 1987.
4. Carroll JE. *Muscle disorders in the newborn.* In Roberton NRC (ed): *Textbook of Neonatogy.* 1st ed. London, Churchill Livingstone, 1986; pp. 586–89.

17

Acute Renal Failure

CHIU Man-chun

1. Normal values[1]

1.1 Plasma creatinine

95% Confidence Limits of Plasma Creatinine in Neonates

Gestation (weeks)	Age (days)				
	2	7	14	21	28
28	40–220	23–145	18–118	16–104	15–95
30	30–192	20–132	17–107	15–95	13–87
32	27–175	19–119	15–97	14–86	12–78
34	24–158	17–109	14–88	12–78	11–71
36	23–143	16–98	12–80	11–71	10–64
38	20–130	13–89	12–72	10–64	9–59
40	18–118	13–81	10–66	9–57	9–53

1.2 GFR

"Normal" GFR in Neonates and Infants (ml/min/1.73 m^2)

Gestational age (weeks)	Mean ± SD	
	1 week	2–8 weeks
25–28	11.0 ± 5.4	15.5 ± 6.2
29–34	15.3 ± 5.6	28.7 ± 13.8
38–42	40.6 ± 14.8	65.0 ± 24.8

❏ From PCr and Length

$$GFR = \frac{K \times L}{PCr}$$

GFR: ml/min/1.73 m^2
L: body length (cm)
PCr: plasma creatinine (mmol)
K: 29 for LBW
40 for full term

2. Causes[2]

Major Causes of Renal Failure in the Neonate

Prerenal	Postrenal obstruction	Intrinsic renal failure
Hypotension caused by	*Urethral obstruction*	*Congenital abnormalities*
• Septic shock	• Posterior urethral	• Cystic dysplasia
• Maternal antepartum	valves	• Hypoplasia
haemorrhage	• Imperforate prepuce	• Agenesis
• Twin-to-twin	• Urethral stricture	• Polycystic kidneys
haemorrhage	• Urethral diverticulum	
• Neonatal haemorrhage	• Megaurethra	*Inflammatory*
• Cardiac surgery	• Ureterocele	• Congenital syphilis or
	• Ureteropelvic or	toxoplasmosis
Congestive heart failure	ureterovesical	• Pyelonephritis
	obstruction	
Asphyxia neonatorum		*Vascular*
	Extrinsic tumors	• Venous thrombosis
Dehydration	• Compressing bladder	• Cortical necrosis
	outlet	• Arterial thrombosis
		• Disseminated intra-
	Neurogenic bladder	vascular coagulation
		Acute tubular necrosis
		• Perinatal asphyxia
		• Dehydration
		• Shock
		• Nephrotoxins

3. Pre-renal vs intrinsic renal failure[3]

Laboratory Indices for Prerenal versus Intrinsic ARF
(Indices for neonates [> 32 weeks] are given in parentheses)

Index[a]	Prerenal	Intrinsic renal
Urine volume	Low (variable)	Low/variable (variable)
(U/P)Cr	> 40 (> 30)	< 20 (< 10)
(U/P)urea	> 8 (> 30)	< 3 (< 6)
Specific gravity	> 1.020 (> 1.015)	< 1.010 (< 1.010)
Urine Osmolality (mosm)	> 500 (> 400)	< 350 (< 400)
(U/P)osm	> 1.3 (> 2.0)	~ 1.0 (< 1.0)
UNa (mmol/l)	< 20 (< 30)	> 40 (> 70)
UCr/PCr	> 40 (> 20)	10–20 (10–20)
FENa% = (U/P)Na/(U/P) Cr × 100	< 1 (< 2.5)	> 3 (> 10)
RFI = UNa/(U/P)Cr	< 1 (< 1.2)	> 3 (> 6.0)

[a]U: urinary;   P: plasma;   Cr: creatinine;   Osm: osmolality;   Na: sodium;
FENa: fractional excretion of sodium;   RFI: renal failure index

4. Approach[4]

4.1 History: family history, oligohydramnios, asphyxia

4.2 Physical examination: abdominal masses, congenital anomalies, respiratory distress, cardiovascular instability, seizures

4.3 Nephrosonogram, perhaps vesicoureterogram and radionuclide scan; if urinary tract obstruction, sterile bladder catheterization

4.4 Blood for complete blood count, electrolytes, BUN, creatinine, osmolality

4.5 Urinalysis, sodium, creatinine, osmolality

4.6 If no obstruction or fluid overload: 0.9 per cent NaCl, 20 ml/kg over 60 to 90 minutes + frusemide 1 mg/kg

4.7 If RFI, FENa, and U/P osmolality suggest prerenal oliguria, repeat fluid challenge if urine output has not increased.

4.8 If diagnostic studies suggest intrinsic renal failure, reduce fluid intake, monitor weight, electrolytes, BUN, creatinine, calcium, phosphorus.

5. Management

5.1 Conservative during acute phase

Fluid overload	Fluid restriction to insensible loss (20–30 ml/kg) + urine output Frusemide 1–5 mg/kg iv

Hyponatraemia
  asymptomatic
  symptomatic
	Fluid restriction
	3% NaCl 6 ml/kg over 1–2 hours

Hyperkalaemia	Ca resonium 1 gm/kg/d oral or rectal Ca gluconate (10%) 0.5 ml/kg iv $NaHCO_3$ 2–3 mmol/kg Insulin 0.1 u/kg and glucose 0.3 g/kg iv
Hypocalcaemia	Ca gluconate (10%) 0.5 ml/kg iv
Hyperphosphataemia	$Al(OH)_3$ 20 mg/kg tid orally or $CaCO_3$ 20–50 mg/kg/day
Hypomagnesaemia	$MgSO_4$ (50%) 0.1 ml/kg im
Acidosis	$NaHCO_3$ 1–3 mmol/kg iv
Hypertension	Hydrallazine 0.2 mg/kg im or iv every 4–8 hours; Fluid restriction, diuretics

5.2 Renal replacement therapy

❏ Indications:

  Not absolute, but to be considered when:

  - Hyperkalaemia persistent and severe, not corrected by conservative measures;
  - Severe and persistent acidosis;
  - Fluid overload with hypertension, congestive heart failure unresponsive to medical treatment;
  - Severe catabolism with rapidly rising urea;
  - Uraemic encephalopathy, pericarditis, pleuritis.

❏ Modalities:

  - Peritoneal dialysis
    Method of choice in most situations. Catheter can be inserted in left side of abdomen instead of the usual subumbilical site for longer accommodation of the catheter and avoiding the bladder which is more abdominal than pelvic in neonates.
  - Continuous Arteriovenous Haemofiltration, CAVH
    Indicated for severe fluid overload with congestive heart failure; and also when abdominal surgery or NEC renders peritoneal dialysis unsuitable.
    Anticoagulation: Initial heparin 20 units/kg iv bolus followed by heparin 7–10 units/kg/hr to keep ACT (activated clotting time) 100–120 seconds, or PTT twice that of normal.

- Continuous Arteriovenous Haemodialysis, CAVHD
  It is used in the treatment of catabolic acute renal failure with hypervolaemia and uraemia requiring high solute clearance not achievable with standard CAVH.

Figure 1. CAVH

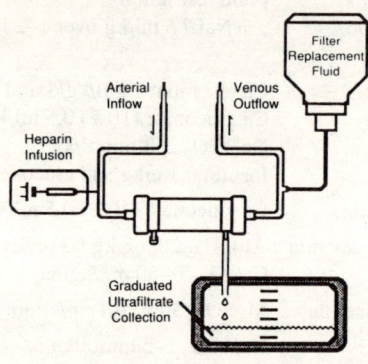

Figure 2. CAVHD

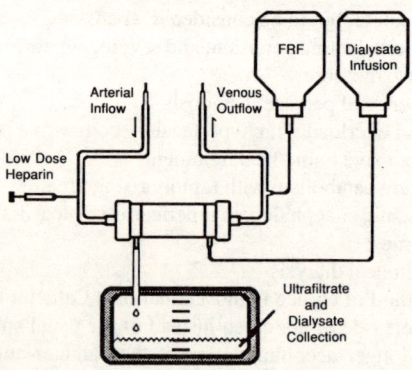

Volume Replacement Fluid = TFC – DI + OFO – TIV ± DFB
DI: Dialysate infusion rate
OFO: Other fluid output
TIV: Total IV fluid
DFB: Desired fluid balance

References

1. Schwartz GJ, Brion LP, Spitzer A. *The use of plasma creatinine concentration for estimating glomerular filtration rate in infants, children and adolescents.* Pediatr Clin North Am 1987; 34(3):575.

2. Rahman N, Boinear FG, Lewy JE. *Renal failure in the perinatal period.* Clin Perinatol 1981; 8:241–50.

3. Kon V, Ichikawa I. *Acute Renal Failure.* In Ichikawa I (ed): *Pediatric Textbook of Fluids and Electrolytes.* Baltimore, Williams & Wilkins, 1990; pp. 283–90.

4. Engle WD. *Evaluation of renal function and acute renal failure in the neonate.* Pediatr Clin North Am 1986; 33(1):141.

5. Mault JR, Dirkes SM, Swartz RD et al. *Continuous Hemofiltration: A Reference Guide for SCUF, CAVH, and CAVHD.* University of Michigan, 1988.

18

Metabolic and Endocrine Disorders

HUEN Kwai-fun

1. Infant of diabetic mother

1.1 Clinical features
❑ Large for gestational age.
❑ Increased incidence of congenital abnormalities especially congenital heart disease and sacral anomalies.
❑ Prone to:
  • hypoglycaemia
  • hyaline membrane disease
  • hypocalcaemia
  • hyperbilirubinaemia
  • polycythaemia

1.2 Treatment
❑ Monitor blood glucose ("dextrostix" or "haemoglucostix") at 0, 1, 2, 4, and 6 hours after birth.
❑ Early enteral feeding within 1–2 hours of delivery.
  • IV glucose at rate of 4–8 mg/kg/min — increase rate if necessary to maintain blood glucose > 2.5 mmol/l.
  • Symptomatic hypoglycaemia: bolus glucose injection 0.2–0.5 g/kg,* followed by a glucose infusion.
❑ Persistent hypoglycaemia despite glucose infusion: IM glucagon 0.03–0.1 mg/kg.
❑ If hypoglycaemia still persists: hydrocortisone 5 mg/kg IV 12 hourly.
❑ Increase enteral feed and decrease parenteral glucose after blood glucose has become stable for 12 hours.

---

* Reactive hypoglycaemia may follow fast bolus glucose infusion.

2. Neonatal hypothyroidism

Refer to Chapter 26.

3. Neonatal hyperthyroidism

3.1 Clinical features
❏ Maternal history of Graves' disease or other autoimmune thyroid diseases.
❏ History of foetal tachycardia.
❏ Most significant signs and symptoms are in the cardiovascular system: tachycardia, cardiac failure and arrhythmia.
❏ Poor weight gain despite good appetite, irritability, exophthalmos and occasionally goitre.

3.2 Investigation
T4, TSI (Thyroid stimulating immunoglobulins)

3.3. Management
❏ Antithyroid drugs:
  • Propylthiouracil 10 mg/kg/day every 8 hours or
  • Carbimazole 1.0 mg/kg/day every 8 hours
  (*Note*: Continue treatment for several weeks and then cautiously wean off.)
❏ For severe cases, add propranolol 2 mg/kg/day 6–8 hourly ± Lugol's iodine solution 1 drop 8-hourly ± treatment for heart failure.

4. Hypopituitarism

4.1 Clinical features
❏ Hypoglycaemia: exclude adrenocortical insufficiency or pan-hypopituitarism.
❏ Newborn males: micropenis with hypoglycaemia — must exclude hypopituitarism in such cases.

4.2 Investigation
❏ Glucagon stimulation test for GH secretion
❏ TRH stimulation test for TSH secretion
❏ LHRH stimulation test for LH/FSH secretion
❏ CRH stimulation test for ACTH secretion
❏ CT scan brain and pituitary for cause of panhypopituitarism

4.3 Management
❏ Hydrocortisone 20–25 mg/m^2/day
❏ T4 100 μg/m^2/day (or 10 μg/kg/day)

❑ GH replacement seldom necessary in first year of life but may be useful in controlling intractable hypoglycaemia.

5. Hypocalcaemia

5.1. Definition
Total serum calcium < 1.75 mmol/l and/or ionized calcium < 0.625 mmol/l

5.2 Clinical features
❑ Irritability, tremors, twitching and seizures, lethargy, poor feeding and vomiting
❑ Prolonged QTc in ECG

5.3 Investigation
❑ Serum calcium, phosphate, alkaline phosphatase and magnesium
❑ Maternal serum calcium
❑ Serum PTH (parathyroid hormone)
❑ CXR for thymic aplasia
❑ Urine for calcium/creatinine, glucose and amino acid, and phosphate excretion
❑ Vitamin D metabolites assessment as indicated

5.4 Treatment
❑ Slow IV injection 10% calcium gluconate (dose: 2 ml/kg) with continuous ECG and heart rate monitoring, followed by continuous infusion 5–8 ml/kg/day and monitor serum calcium level.
❑ Oral calcium gluconate 200–500 mg/kg/day.
❑ Add Vitamin D analogue if hypocalcaemia persists, e.g. 1,25-dihydroxycholecalciferol 0.25–2 µg/day in 1–4 doses (avoid use of vitamin D if vitamin D metabolites have to be assessed).
❑ Further treatment depends on cause of hypocalcaemia.

6. Hypercalcaemia

6.1 Definition
Serum calcium > 2.75 mmol/l

6.2 Clinical features
Hypotonia, weakness, irritability, poor feeding, weight loss, constipation, vomiting, polyuria, polydipsia

6.3 Investigation
❑ Calcium, phosphate, alkaline phosphatase
❑ Urine calcium/creatinine ratio
❑ Serum PTH level

6.4 Treatment
❑ Generous hydration and frusemide

❏ Hydrocortisone 1 mg/kg 6-hourly
❏ Calcitonin 10 u/kg IV, may be repeated 4-hourly

7. Congenital adrenal hyperplasia

7.1 95% due to 21-hydroxylase deficiency

7.2 **Clinical features**
❏ Male: hyperpigmentation of scrotum and penis
❏ Female: ambiguous genitalia, virilization
❏ For salt-losing type: vomiting, anorexia, diarrhoea, severe dehydration and shock

7.3 **Investigation**
❏ Electrolytes, blood gas, blood glucose
❏ Karyotype
❏ Plasma 17α-hydroxyprogesterone (17αHP)

7.4 **Treatment**
❏ Rehydration with plasma or saline solution
❏ Hydrocortisone 20–25 mg/m^2/day in 2–3 divided doses
❏ Fludrocortisone 50–200 µg/day
❏ Salt supplement 2–4 g NaCl/day
❏ Surgical correction of genital abnormality

8. Inborn errors of metabolism (IEM)

8.1 **Clinical symptomatology**
❏ Symptoms indicating possibility of an IEM (one or all)
  • Infant becomes acutely ill after a period (hours–weeks) of normal behaviour and feeding.
  • Neonate or infant with seizures and/or hypotonia especially if seizures are intractable.
  • Neonate or infant with an unusual odour.
❏ Symptoms indicating strong probability of an IEM, particularly when coupled with the above symptoms
  • Persistent or recurrent vomiting
  • Failure to thrive (failure to gain weight or weight loss)
  • Apnoea or respiratory distress (tachypnoea)
  • Jaundice or hepatomegaly
  • Lethargy
  • Coma (particularly intermittent)
  • Unexplained haemorrhage
  • Family history of neonatal deaths, or of similar illness especially in siblings
  • Parental consanguinity

  - Sepsis (particularly Escherichia coli)

8.2 Physical anomalies associated with acute onset inborn errors of metabolism (IEM)

Anomaly	Possible IEM
Ambiguous genitalia	Congenital adrenal hyperplasia
Hair and/or skin problems (alopecia, dermatitis)	Multiple carboxylase deficiency, biotinidase deficiency, argininosuccinic aciduria
Structural brain abnormalities (agenesis of corpus callosum, cortical cysts)	Pyruvate dehydrogenase deficiency
Macrocephaly	Glutaric aciduria, type I
Renal cysts, facial dysmorphism	Glutaric aciduria, type II; Zellweger syndrome
Facial dysmorphism	Peroxisomal disorders, (Zellweger syndrome)
Cataract	Galactosaemia, Łowe syndrome
Retinopathy	Peroxisomal disorders
Lens dislocation, seizures	Sulfite oxidase deficiency, Molybdenum cofactor deficiency
Facial dysmorphism congenital heart disease vertebral anomalies	3-OH-isobutyric CoA deacylase deficiency

8.3 Laboratory tests
  - ❏ Commonly available tests
    - Complete blood count with phase platelet count (leukopenia, thrombocytopenia, anaemia)
    - Urinalysis (ferric chloride, pH, ketones, clinistix [glucose], clinitest reducing substance)
    - Arterial blood gases (acidosis, alkalosis)
    - Serum electrolytes (anion gap, hyperchloraemia)
    - Blood glucose (hypoglycaemia)
    - Blood ammonia (hyperammonaemia)
    - Lactic and pyruvic acids (from serum and cerebrospinal fluid, obtained simultaneously)
    - Uric acid (hyperuricaemia)
    - Liver function studies (eg, elevated transaminases, decreased clotting factors)
    - Blood urea nitrogen (low values)
  - ❏ Available specific tests for IEM (on-site or from experienced metabolic reference laboratory)
    - Quantitative amino acids (plasma, by column chromatography)

- Quantitative amino acids (cerebrospinal fluid, by column chromatography)
- Organic acids (urine, gas chromatography and/or gas chromatography-mass spectrometry)
- Carnitine (plasma total and free)
- Very long chain fatty acids (serum)
- β-OH-butyrate (serum)
- Free fatty acids (serum)
- Quantitative orotic acid (urine)
- Galactose-1-phosphouridyl transferase (red blood cell)

❏ More limited available tests, usually requiring clinical genetics/ metabolic professional interface
- Enzyme determination
  Cultured skin fibroblasts
  Leukocytes
  Liver
- Molecular DNA studies (index case)
  Whole blood leukocytes
  Skin fibroblasts
  Liver or other tissue

8.4 Nonacidotic, nonhyperammonaemic features

❏ Neurologic features predominant: seizures, hypotonia, optic abnormality
- Glycine encephalopathy (nonketotic hyperglycinaemia)
- Pyridoxine-responsive seizures
- Sulfite oxidase/xanthine oxidase deficiency
- Peroxisomal disorders (Zellweger syndrome, neonatal adrenoleukodystrophy, infantile refsum disease)

❏ Jaundice prominent
- Galactosaemia
- Hereditary fructose intolerance
- Menkes kinky hair syndrome
- α1-antitrypsin deficiency

❏ Hypoglycaemia (nonketotic)
- Fatty acid oxidation defects (MCAD, LCAD, carnitine palmityl transferase, infantile form)

❏ Cardiomegaly
- Glycogen storage disease (type II phosphorylase kinase b deficiency)
- Fatty acid oxidation defects (LCAD)

❏ Hepatomegaly (fatty liver)
- Fatty acid oxidation defects (MCAD, LCAD)

❏ Skeletal muscle weakness
- Fatty acid oxidation defects (LCAD, SCAD, multiple acyl-CoA dehydrogenase)

8.5 Flow chart

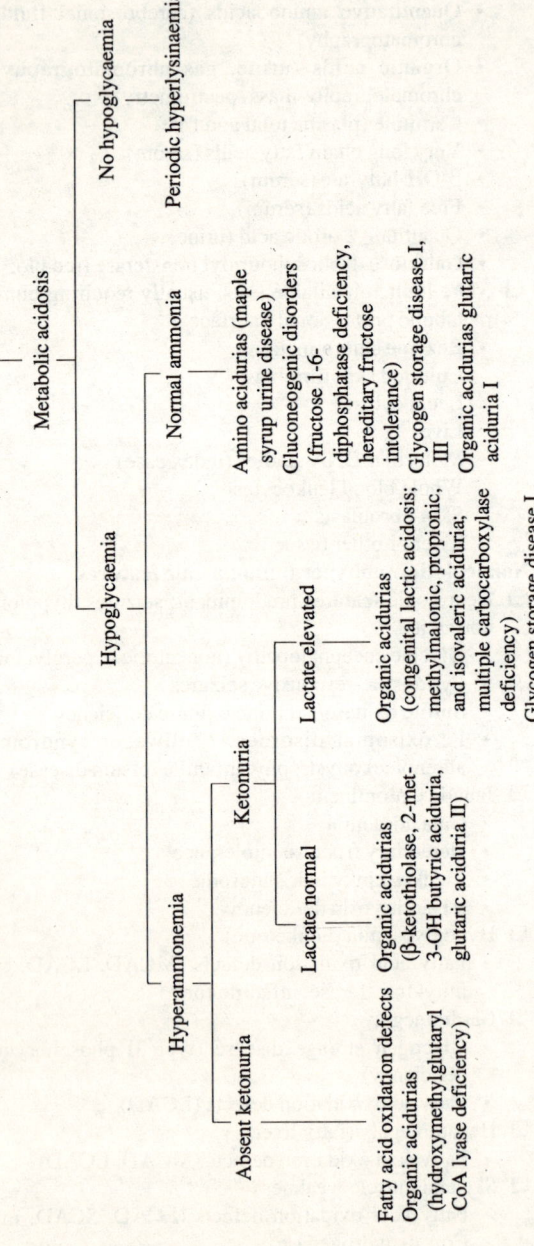

Inborn error of metabolism

Metabolic acidosis

- Hypoglycaemia
  - Hyperammonemia
    - Absent ketonuria

      Fatty acid oxidation defects
      Organic acidurias (hydroxymethylglutaryl CoA lyase deficiency)

    - Ketonuria
      - Lactate normal

        Organic acidurias (β-ketothiolase, 2-met-3-OH butyric aciduria, glutaric aciduria II)

      - Lactate elevated

        Organic acidurias (congenital lactic acidosis; methylmalonic, propionic, and isovaleric aciduria; multiple carbocarboxylase deficiency)
        Glycogen storage disease I

  - Normal ammonia

    Amino acidurias (maple syrup urine disease)
    Gluconeogenic disorders (fructose 1-6-diphosphatase deficiency, hereditary fructose intolerance)
    Glycogen storage disease I, III
    Organic acidurias glutaric aciduria I

- No hypoglycaemia

  Periodic hyperlysinaemia

Inborn error of metabolism
No metabolic acidosis
No hypoglycaemia

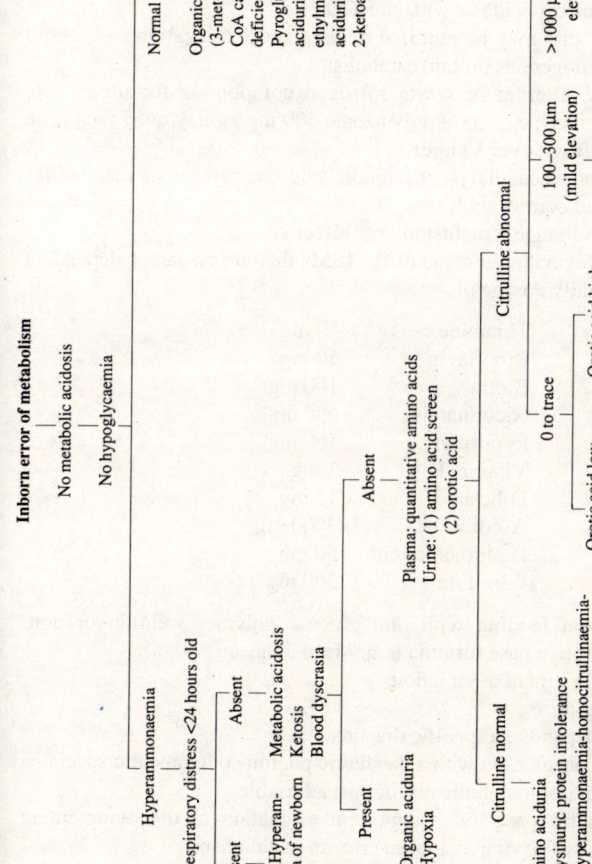

Normal ammonia → Organic aciduria (3-met-crotonyl CoA carboxylase deficiency, Pyroglutamic aciduria, ethylmalonic aciduria, possible 2-ketoadipic)

Hyperammonaemia
- Respiratory distress <24 hours old
  - Present → Transient Hyperammonaemia of newborn
  - Absent
    - Metabolic acidosis / Ketosis / Blood dyscrasia
      - Present → Organic aciduria / Hypoxia
      - Absent → Plasma: quantitative amino acids; Urine: (1) amino acid screen (2) orotic acid
        - Citrulline normal → Amino aciduria / Lysinuric protein intolerance / Hyperammonaemia-homocitrullinaemia-hyperornithinaemia (H-H-H) / Periodic hyperlysinaemia
        - Citrulline abnormal
          - 0 to trace
            - Orotic acid low → Carbamyl phosphate synthetase or N-acetyl glutamate synthetase
            - Orotic acid high → Ornithine carbamyl transferase
          - 100–300 µm (mild elevation) → ASA anhydrides urine/plasma → Argininosuccinic acidaemia
          - >1000 µm (marked elevation) → Citrullinaemia

8.6   Treatment
❏ Acute
  • Stop milk intake.
  • Give IV D10/D15 and essential amino acids 0.5–0.75g/kg/day
    (arginine supplement essential for urea cycle defects) to en-
    sure adequate caloric intake (> 70 cal/kg/day) and to prevent
    catabolism.
  • Correct acidosis with $NaHCO_3$.
  • Insulin may be required to aid glucose utilization and inhibit
    endogenous protein catabolism.
  • IV alternative waste nitrogen compounds for urea cycle
    defects, e.g. sodium benzoate 500 mg/kg/day q6h: each dose
    infused over ½ hour.
  • Peritoneal dialysis/haemodialysis to remove toxic metabolites
    and correct acidosis.
  • Exchange transfusion: less effective.
  • Megavitamin cocktail for IEMs that are co-factor dependent
    (daily dosages):

Thiamine	50 mg
Riboflavin	50 mg
Biotin	100 mg
Nicotinamide	600 mg
Pyridoxine	100 mg
Vitamin B_{12}	1 mg
Folic acid	15 mg
Ascorbic acid	3000 mg
Pantothenic acid	50 mg
Carnitine	300 mg

  • Oral feeding with oral glucose polymer solution or non-
    protein base formula (e.g. Mead-Johnson 80056).
  • Treatment of infections.
❏ Long-term
  • Depends on specific diagnosis
  • Consultation with a paediatric pharmacotherapeutic specialist
    and/or metabolic nutritionist advisable

8.7   Protocol for pre- and postmortem evaluation of the dying infant suspected of having an inborn error of metabolism
❏ Blood (10–20 mL)
  • Serum and plasma: separate, freeze in aliquots of 1–2 mL at
    −20°C (for quantitative amino acids, carnitine, ketone bodies,
    etc.)
  • Erythrocyte and leukocyte fractions
    – Erythrocytes: Refrigerate at 4°C (for selected enzyme,
      peroxisomal studies)

    – Leukocytes: Freeze a fraction at –20°C (for possible enzyme/DNA studies)
❏ Urine (20–30 mL)
    Store in 5-mL aliquots at –20°C (for organic acids, amino acid screening, orotic acid, and others)
❏ Vitreous humour (chemistries)
❏ Skin (3–4 mm)
- Full-thickness skin biopsy taken sterilely (cleanse site with alcohol, not iodine, and use sterile instrument)
- Store in sterile culture medium (or sterile isotonic saline with glucose) and contact fibroblast tissue culture laboratory immediately.
❏ Tissue biopsies (liver, heart, muscle, brain, etc.)
- Needle biopsy or immediate autopsy samples of tissue (quick frozen with dry ice [aluminium foil on dry ice]); transfer to pre-labelled tightly stoppered tubes; freeze at –20°C.
- Light and electron microscopy; particularly of liver, heart, and skeletal muscle tissue

9. Clinical approach to problem of ambiguous external genitalia

9.1 Review the pre- and perinatal history for clues such as exposure to drugs.

9.2 Re-examine the infant for pigmentation, palpable gonads or other relevant features.

9.3 Keep mother and baby in hospital for at least 2 weeks and check regularly for hypoglycaemia, hyperkalaemia, salt loss or other metabolic problems.

9.4 Obtain a karyotype as quickly as possible.

9.5 Chemical investigation: tailored according to the clinical findings.

9.6 Anatomical investigation: genitogram ± ultrasound and CT.

References

1. Aynsley-Green A, Soltesz G. *Disorders of Blood Glucose Homeostasis in the Neonate*. In Roberton NRC (ed): *Textbook of Neonatology*. 1st ed. Churchill Livingstone, 1986; pp. 605–22.

2. Barnes ND. *Endocrine Disorders*. In Roberton NRC (ed): *Textbook of Neonatology*. 1st ed. Churchill Livingstone, 1986; pp. 623–43.

3. Danks DM, Brown GK. *Inborn Errors of Metabolism in the Neonate*. In Roberton NRC (ed): *Textbook of Neonatology*. 1st ed. Churchill Livingstone, 1986; pp. 644–58.

4.   Brook CGD. *Clinical Paediatric Endocrinology*. 2nd ed. Blackwell Scientific Publication, 1989.

5.   *Paediatrics in Review*, 1990; 11(7):205–16.

6.   Greene CL. *Inborn errors of metabolism and Reye Syndrome: Differential diagnosis*. J Pediatr 1988; 113:156–59.

7.   Hudak ML. *Differentiation of transient hyperammonaemia of the newborn and urea cycle enzyme defects*. J Pediatr 1985; 107:712–19.

19

Acute Gastrointestinal Diseases

Ellen PN LEUNG

Common presentation of gastrointestinal (GI) disorders
Upper GI diseases:
- Vomiting or increased gastric residue
- Coffee ground vomitus
- Abdominal distension

Lower GI diseases:
- Diarrhoea
- Delayed passage of meconium
- Blood per rectum
- Abdominal distension

1. Vomiting

1.1 Definition

Forceful ejection of stomach contents.

1.2 Timing and nature of vomitus and its clinical correlation

- ❑ Vomiting right after the first feed, or associated with abdominal distension: may be a manifestation of an anatomical obstruction.
- ❑ Bile stained vomitus: a surgical emergency until proven otherwise.
- ❑ Frothy vomitus: suggestive of oesophageal atresia.
- ❑ Blood stained vomitus: possibilities:
  - Swallowed maternal blood
  - Haemorrhagic disease of newborn
  - Feeding tube trauma
  - Stress ulcers
  - Hiatus hernia, etc.
- ❑ Failure to gain weight: suggests a more chronic process.

1.3 Causes of vomiting in the newborn

❏ Benign — feeding disorders, physiological immaturity of gastroesophageal junction.

❏ Surgical condition (partial/complete GI obstruction or other structural abnormalities), e.g.:
oesophageal atresia, ileal atresia, duodenal atresia, duodenal bands, strangulated hernia, malrotation, volvulus, Hirschsprung's disease, hiatus hernia, meconium plug, etc.

❏ Functional ileus:
- Electrolyte imbalance, extreme preterm, severe RDS.
- Infection: UTI, septicaemia, gastroenteritis, NEC.

❏ Intracranial lesions, e.g.:
cerebral oedema, intracranial bleeding, meningitis, subdural effusion.

❏ Miscellaneous: adrenal insufficiency, uraemia, inborn error of metabolism, drugs (digoxin, antibiotics, narcotic withdrawal).

1.4 Management

❏ Detailed examination to exclude underlying causes and assess state of hydration.

❏ Investigation (guided by clinical suspicion):
Electrolytes
Blood gas
Blood glucose
Urinalysis
Bacterial cultures (MSU, blood, CSF)
Abdominal X-rays

❏ Treatment:
Correct dehydration.
Suspend feeding if vomiting severe and persistent.
Further management will depend on underlying cause.

1.5 Management for gastroesophageal reflux

Small frequent feeds
Frequent "burping"
Prop up patient in prone position
Thickened feeds with rice cereal
± drugs (antacids, dopamine antagonist, prepulsid)
Rarely surgical intervention

2. Diarrhoea

2.1 Definition

Frequent passage of watery or loose stool.
(*Note*: up to 8 motions per day is probably normal for bottle-fed; breast-fed infants may have as many as 16 stools/day.)

2.2 Causes of neonatal diarrhoea
- ❏ Sepsis:
    - Gastroenteritis
    - NEC
    - UTI
    - Septicaemia
- ❏ Maternal drug effect (especially in breast-fed infant):
    - Laxatives
    - Drug addiction
- ❏ Drug treatment:
    - Penicillin
    - Iron
    - Oral calcium
    - Xanthines
    - Phototherapy also cause loose stool
- ❏ Miscellaneous:
    - Milk allergy
    - Disaccharide intolerance
    - Cystic fibrosis
    - Thyrotoxicosis
    - CAH, etc.

2.3 Clinical evaluation
- ❏ Hx:
    - Technique in preparation of nipple or formula milk, bowel status of other family members
    - Associated symptoms (fever, vomiting)
    - Birth weight and current body weight
- ❏ P/E:
    - Hydration status
    - Stool inspection
    - Sore buttock
- ❏ Ix:
    - Stool — microscopy, pH, reducing substance
    - Urine — urinalysis, S.G. and osmolality
    - Blood — electrolytes, urea, blood gas

2.4 Gastroenteritis
- ❏ The main dangers are dehydration, septicaemia and cross infection.
- ❏ Common pathogens:
    - Toxigenic E. Coli
    - Shigella
    - Salmonella
    - Campylobacter jejuni
    - Rotavirus
    - Astrovirus

(*Note*: In over 60% of cases, no organism can be isolated.)

❑ Management
  • Rehydration
  Mild:
    – Suspend milk feeds
    – Replace with electrolyte solution
  Severe:
    Nil by mouth
    – The mainstay of Rx is assurance of adequate hydration and electrolyte balance.
    – IV fluid: correct deficit
             adequate maintenance
             adjust for current loss
    – Hypernatraemic dehydration (serum Na+ > 150 mmol/l):
             Slow rehydration mandatory, correct deficit in 48–72 hrs
             Plasma, NS, ½:½ solution, etc. are the preferred replacement fluids
    – Antibiotic has no place unless septicaemia suspected
    – Antidiarrhoeal agents not recommended
  • Isolation of infected infants:
    – Cohort babies with +ve culture
    – Cohort separately babies with diarrhoea pending culture result
    – Meticulous hand washing is essential
    – Careful disposal of excretor
  • Recurrence of diarrhoea after reintroduction of milk feed:
    – If stools are acidic (pH < 5.5) and (+)ve for reducing substance: Suspect secondary lactose intolerance and try lactose-free or low lactose formulae.

2.5 Necrotizing enterocolitis
(Please refer to Chapter 20.)

3. Abdominal distension

3.1 Causes
  ❑ Organic (majority related to surgical condition):
       Gut atresia/stenosis/webs
       Malrotation
       Volvulus
       Meconium plug
       Hirschsprung's disease, etc.

❏ Functional (paralytic ileus):
  Sepsis
  Asphyxia neonatorum
  Drug withdrawal
  NEC
  Hypothyroidism, etc.
❏ Others:
  Pneumoperitoneum
  Bilateral pneumothoraces
  Ascites
  Renal masses, etc.

3.2 Management

(Refer to Chapter 21.)

❏ A careful history and examination will usually distinguish between the causes of functional obstruction from true mechanical
obstruction.

4. Blood in stool

4.1 Causes

❏ Swallowed maternal blood
❏ Local:
  Anal fissure
  NEC
  Malrotation of gut
  Duplication of bowel
  Rarely gastroenteritis (rotavirus, astrovirus, campylobacter,
    shigella), etc.
❏ General: Haemorrhagic disease of the newborn

4.2 Management

❏ Volume resuscitation is the first priority if bleeding is profuse and
associated with vascular instability. (Save samples of blood for
investigation before blood transfusion.)
❏ Find out the possible cause of bleeding:
  Detailed Hx and P/E to look out for bleeding tendency.
  Relevant Ix:
    sepsis screen
    stool culture (bacterial, viral)
    Apt's test
    AXR
    Ix for bleeding diathesis, etc.
❏ Further management would be directed to primary cause.

References

1. Avery ME, First LR. *Pediatric Medicine*. 1st ed. Baltimore, Williams & Wilkins, 1989.
2. Halliday HL, McClure G, Reid M. *Handbook of Neonatal Intensive Care*. 2nd ed. England, Bailliere Tindall, 1985.
3. Roberton NRC. *A Manual of Neonatal Intensive Care*. 2nd ed. Great Britain, Edward Arnold, 1986.
4. Roberton NRC. *Textbook of Neonatology*. 1st ed. Great Britain, Churchill Livingstone, 1986.

20
Surgical Emergencies

NG Wai-dat

A. NEONATAL INTESTINAL OBSTRUCTION

1. Causes

1.1 "Medical" causes for adynamic ileus
- ❑ Sepsis/septicaemia
- ❑ Anoxia/asphyxia/respiratory distress
- ❑ Cerebral birth trauma
- ❑ Adrenal insufficiency/hypothyroidism
- ❑ Functional immaturity (transient obstruction in premature babies, meconium plug syndrome, small left colon syndrome)

1.2 "Surgical" causes
- ❑ Mechanical obstruction:
  Atresia (types I–IV), stenosis, inspissated meconium, congenital bands, volvulus, hernia, intussusception, duplication
- ❑ Functional obstruction:
  Hirschsprung's disease, neuronal dysplasia
- ❑ Paralytic ileus (secondary to peritonitis):
  Gastrointestinal perforation (unrelieved obstruction, bowel gangrene, iatrogenic, idiopathic), enterocolitis

2. Clinical suspicion

2.1 Antenatal
- ❑ Maternal hydramnios (20–40%)
- ❑ Sonograms

2.2 Postnatal

❏ Abdominal distension, usually 12–24 hours after birth (the lower the obstruction, the more prominent the distension)
❏ Absent/delayed (> 24 hours) passage of meconium (small amounts may be passed even with complete obstruction)
❏ Bilious vomiting/greenish gastric aspirate: this is the only absolute sign of intestinal obstruction.
❏ Other features include abdominal mass (see Figs. 1 and 2) and gastrointestinal bleeding.

3. Examination

3.1 Fontanelle fullness/skin turgor; colour of fingers and toes; temperature.
3.2 Abdominal distension; tenderness; mass.
3.3 Rectal examination: anorectal anomalies, blood on glove, Hirschsprung's disease (gripping and gush of meconium on withdrawal).

4. Radiographic investigations

4.1 Start with supine and erect/decubitus films
❏ Multiple fluid levels in thumb-sized bowel loops indicate complete obstruction.
❏ Uniformly dilated loops of gas-filled bowel but without definite fluid levels suggest "medical" ileus.
❏ Peritoneal calcification signifies meconium peritonitis resulting from early intrauterine bowel perforation.
❏ Hirschsprung's disease is evidenced by generalized gaseous distension with absence of rectal gas; rarely is the transitional zone outlined by gas.
❏ Intussusception is evidenced by interruption of the barium flow and "coil-spring" sign.
❏ Midgut volvulus is often characterized by gasless abdomen.
❏ Perforation with pneumoperitoneum or evidence of enterocolitis may be recognized in advanced cases.
4.2 Barium enema
❏ To distinguish between small and large bowel distension.
❏ To determine if the colon is used or unused (microcolon indicates proximal complete obstruction occurring well before birth).
❏ To locate the position of the caecum in regard to intestinal rotation and fixation. "Bird's beak" appearance is characteristic at the site of twisting.

Figure 1. Diagnostic Approach to Bile-stained Vomiting in the Newborn

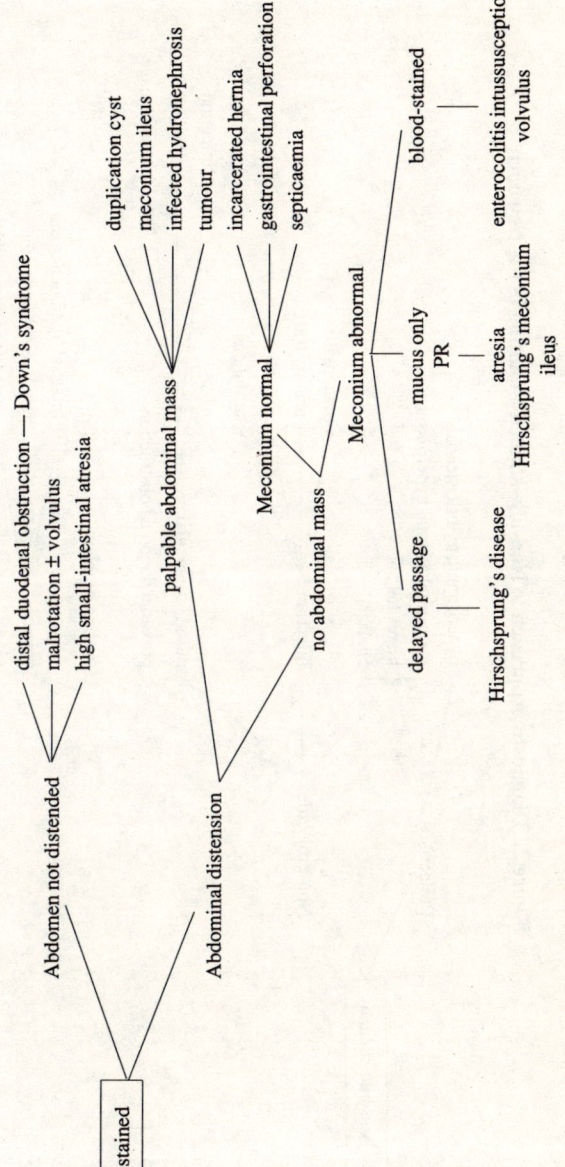

Figure 2. Diagnostic Approach to Non-bile-stained Vomiting in the Newborn

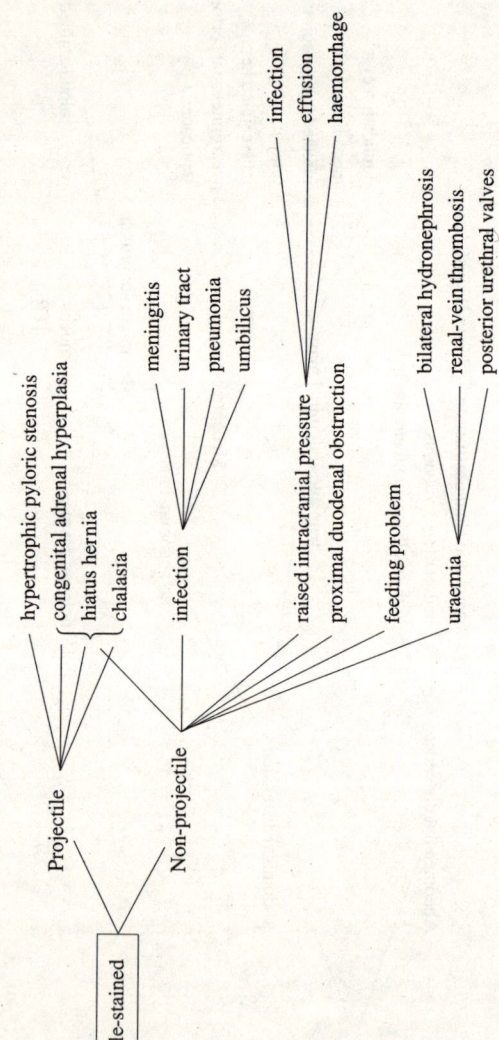

❏ In neonatal Hirschsprung's disease, the cone segment may be ill-defined: A saw-toothed appearance of the proximal spastic colon may be seen. Alternatively, barium retention for 2–3 days may be the only tell-tale evidence.

❏ The study may delineate colonic atresia or small left colon syndrome.

❏ May even be therapeutic for meconium plug syndrome.

4.3 Ba meal and follow through

Indicated for clinically high obstruction. Malrotation may be suggested by abnormal position of the ligament of Treitz and duodenal obstruction.

5. Rectal biopsy

5.1 Indicated if Hirschsprung's disease is suspected.

5.2 Using a suction apparatus, multiple biopsies are taken from rectum at least 1.5 cm above the pectinate line.

5.3 Diagnosis is confirmed by absence of ganglion cells and an abundance of hypertrophic fibres, made prominent by acetylcholinesterase staining. (The latter may be unapparent in the lamina propria in newborns up to 3 months of age.) If biopsy shows decreased or normal ganglion cell population, the case can be regarded as neonatal functional obstruction.

6. Preoperative management

6.1 Nasogastric suction

6.2 Energetic IV fluid therapy (replace gastric juice loss ml by ml)

6.3 Broad-spectrum antibiotics

6.4 Cross-matching.

7. Early operation

Required in cases with:

7.1 Gross and tense abdominal distension

7.2 Well-defined and persistent fluid levels

7.3 Pneumoperitoneum

7.4 Septicaemia secondary to obstruction

7.5 Suspected presence of gangrenous bowels, e.g. abdominal wall oedema and periumbilical erythema

7.6 Histologically confirmed Hirschsprung's disease

B. NECROTIZING ENTEROCOLITIS (NEC)

1. Predisposing factors

1.1 Prematurity
1.2 Respiratory distress/asphyxia
1.3 Symptomatic cardiac anomalies
1.4 Congenital GI anomalies
1.5 Shock/hypothermia/hypoxia
1.6 Polycythaemia/anaemia/thrombocytosis
1.7 ET/umbilical catheterization
1.8 Too much/too fast/hypertonic/milk formula

(*Note*: NEC may occur in term babies without the above risk factors.)

2. Clinical course

2.1 Early or mild
   ❑ Lethargy, bradycardia, temperature instability, recurrent apnoea.
   ❑ Abdominal distension, gastric retention, emesis, passage of loose stools ± red blood.
2.2 Moderate degree
   ❑ Mild metabolic acidosis, mild thrombocytopenia
   ❑ Abdominal tenderness, absent bowel sounds
2.3 Advanced
   ❑ Hypotension, severe apnoea, neutropenia, combined respiratory and metabolic acidosis, DIC
   ❑ Marked abdominal tenderness, oedema/erythema of abdominal wall, localized/fixed mass

3. Radiology

3.1 Because of the non-specific nature of the clinical features, radiology is important for the diagnosis:
   ❑ Multiple gas-filled loops of dilated bowel ± fluid levels: early NEC
   ❑ Foamy cystic/linear pneumatosis, portal vein air, small bowel separation: moderate NEC
   ❑ Marked ascites (peritonitis), fixed dilated loops (necrosis), free gas (perforation): advanced NEC
3.2 Repeat abdominal X-ray AP and left lateral decubitus every 8–12 hours.

4. Management

4.1 For early cases
- ❏ Withhold feeding + nasogastric decompression
- ❏ Monitoring with repeat XR examination, platelet counts, WBC, blood gas
- ❏ Sepsis work up
- ❏ Antibiotics (ampicillin + aminoglycoside) for 4 days pending culture results

4.2 For moderate cases, add:
- ❏ Antibiotics (add metronidazole) to be continued for 1–2 weeks
- ❏ Gammaglobulin may be given
- ❏ Intravenous alimentation
- ❏ On recovery, gradually re-introduce enteral feeding with small volumes of dilute lactose-free formula 2 weeks later. Concurrently, stools are tested for reducing substances and blood. Feedings are discontinued if either tests becomes positive.

4.3 For advanced cases, add:
- ❏ Energetic fluid replacement: 200 ml/kg/day above maintenance requirement
- ❏ Assisted ventilation
- ❏ Inotropic agents: low dose dopamine (2–5 µg/kg/min) (high doses may reduce mesenteric blood flow). If further inotropic support is needed, dobutamine may be added (5–10 µg/kg/min).
- ❏ Timing of surgery is crucial, indications are:
  - Pneumoperitoneum: The only sign invariably indicating bowel necrosis
  - Continued rapid deterioration despite adequate supportive treatment, as judged by a concurrence of multiple criteria (including worsening signs of peritonitis, progressive and uncorrectable ventilatory failure, and falling platelet counts)
  - Presence of evidence of full-thickness necrosis:
    - Fixed tender mass
    - Fixed persistent dilated loops on repeat AXR
    - Abdominal wall oedema/erythema
  - Positive paracentesis, defined as: "Brownish fluid and/or bacteria on gram smear; volume must be over 0.5 ml."
  - Extensive pneumatosis (however, extensive curvilinear pneumatosis on both sides of the abdomen indicates long segment involvement and poor surgical candidate)
  - Failure to improve on medical treatment
  - Ascites:
    For dry tap: Instill 30 ml/kg NS, turn patient from side to side, and withdraw the fluid.

❑ Surgery is aimed at:
- Excision of septic segment(s) and peritoneal toilet, followed by
- Double enterostomy to decompress the proximal bowel and to defunction the distal bowel
- To prevent excessive fluid and salt losses, enterostomy is to be closed early, usually a few weeks later, preceded by barium study excluding strictures.

4.4 For critically ill infants in whom laparotomy would otherwise be indicated: peritoneal drainage via indwelling catheter (usually placed in RLQ under LA in the ward), pending surgery when clinical response is achieved.

C. OESOPHAGEAL ATRESIA

1. Anatomic patterns

Atresia with distal fistula
Atresia without fistula
Fistula without atresia
Atresia with proximal fistula
Atresia with double fistulae

2. Coexistent congenital malformations

Overall (1/2 of the cases)
Cardiac — PDA, VSD (1/3)
GI tract — Imperforate anus, duodenal atresia (1/5)
VATER syndrome
Trisomy 18

3. Different risk groups

Selective surgical management is based upon categorization of infants into different risk groups:

3.1 Category A: Birth weight over 2.5 kg and otherwise well
3.2 Category B
❑ Birth weight 1.8–2.5 kg and well, or
❑ Higher birth weight but moderate pneumonia and other congenital malformations

3.3 Category C
- ❏ Birth weight under 1.8 kg, or
- ❏ Higher birth weight but severe pneumonia and severe congenital malformations

4. Hazards

4.1 Spilling of saliva or feeds into the trachea and lungs from the over-filled upper pouch
4.2 Reflux of acidic gastric juice up the distal oesophagus, traversing the fistula and down the bronchus causing chemical pneumonitis

5. Clinical suspicions

5.1 Regurgitation of saliva, "excessive mucus"
5.2 Attacks of choking, coughing, cyanosis and respiratory distress, all provoked by feeding dating back to first feeding
5.3 Abdominal distension, more evident on coughing or crying
5.4 Pulmonary difficulties/pneumonia, atelectasis (especially RUZ)

6. Diagnosis

Early diagnosis imperative for any delay inevitably leads to progressive pulmonary complication and poor surgical outcome.
6.1 Arrest of a firm rubber catheter at about 10 cm from the lips establishes unequivocally the diagnosis. This simple test should be performed on all newborns suspected to have oesophageal atresia without delay, and certainly before the infant is fed.
6.2 In the X-ray, look for:
- ❏ Coiling up of catheter in superior mediastinum
- ❏ The state of the lungs
- ❏ Thoracic skeletal anomalies
- ❏ Gas pattern in the GI tract signifying the presence/absence of a tracheo-oesophageal fistula

7. Initial management

7.1 Nurse baby at 30 degree head up, semi-prone position to minimize aspiration of gastric secretion into the lungs through the fistula. Baby's position must be changed from side to side to prevent atelectasis, every time after prior aspiration of the upper pouch.

7.2 Frequent oro-tracheal suctioning plus continuous suctioning of upper pouch with a sump catheter.
7.3 Antibiotics
7.4 Treatment regimen:
   ❏ Category A: Immediate operative repair with division of fistula and primary anastomosis
   ❏ Category B: Stabilization (with suction, antibiotics, IV alimentation and gastrostomy drainage) for a few days, then transthoracic repair
   ❏ Category C: Staged repair, start with:
     • Division of fistula, gastrostomy feedings and upper pouch suction, or
     • Pouch suction, gastric division with double gastrostomy, or
     • Pouch suction, IV alimentation and gastrostomy drainage pending elective transthoracic repair

8. Complications and sequelae

8.1 Brassy cough
   Not uncommon and may persist for 12 to 18 months till the trachea has grown sufficiently.
8.2 Oesophageal stricture
   Most improve with bouginage but a few require resection.
8.3 Anastomosis leakage
   Small partial breakdown may heal following proper drainage, while complete dehiscence is treated by cervical oesophagostomy and closure of lower end of the oesophagus.
8.4 Recurrence of tracheo-oesophageal fistula
   Manifested by recurrent chest infections, requiring careful radiographic confirmation. Re-operation is needed.

D. ANORECTAL MALFORMATIONS

1. Types

For practical purposes, two categories are recognized:
1.1 High type/supralevator/rectal lesion: Arrested development occurred above the chief muscle of continence (puborectalis).
1.2 Low type/translevator/anal lesion: Deficient development occurred after traversing the levator sling.

2. External examination

2.1   Male babies
- ❏ Complete absence of anal features with no contraction of external sphincter on stretching the perineum suggests a high lesion.
- ❏ A fistula tract along the perineal raphe usually, but not always indicates anal type.
- ❏ The presence of a well-developed bucket-handle deformity in the anal region often signify covered anus.

2.2   Female babies
- ❏ If some meconium can be passed, observe carefully for:
   "One hole": common cloaca
   "Two holes": rectovaginal fistula
   "Three holes":* ano-or rectovestibular fistula, anterior anus, covered anus
- ❏ If meconium cannot be passed, the anomaly is non-communicating (a rarity), and an invertogram is required.

3. Neurological examination

3.1   Pinprick sensation of the sacral dermatome
3.2   Expressibility of the bladder

4. Imaging studies

4.1   Babygram to look for:
- ❏ Cardiac, vertebral, rib and limb deformities
- ❏ Sacral agenesis, < 3 segments (suspect neurologic defects)

4.2   Ultrasonogram to detect renal anomalies/agenesis
4.3   Invertogram: reliable only when attention is paid to details:
- ❏ Baby should be > 12 hours of age in order to allow gas to pass to the distal end of GI tract.
- ❏ Hold baby vertically upside down for 3 minutes before taking film.
- ❏ Keep baby's hip relatively straight so that the femora do not obscure the pubic bone.
- ❏ A thin paste of barium is smeared on the perineal skin.
- ❏ A true lateral view is mandatory.

---

* Differentiated by the site and size of the orifice, and the direction taken by the probe while probing the orifice.

Interpretation:
- ❏ PC line: A line drawn between the centre of the pubic bone and a point 0.5 cm below the ossified portion of S5
- ❏ "I" point: The forward, lowermost lip of the comma-shaped ischial bone

If the rectal gas ends:
- ❏ Above PC line — high type
- ❏ Between PC line and "I" point — intermediate type
- ❏ Below "I" point — low type
- ❏ Much higher than expected and gas shadow not smooth — repeat

4.4 Sagittal CT and MRI can now provide an accurate picture of anorectal anomalies and their associated fistulae

5. Surgical treatment

5.1 For high and intermediate type: colostomy
5.2 For low type: perineal anoplasty followed by regular sounding

E. EXOMPHALOS AND GASTROSCHISIS

1. Definitions

1.1 Exomphalos minor (hernia of the umbilical cord): umbilical defect < 5 cm; containing only intestine
1.2 Exomphalos major: umbilical defect > 5 cm, sac > 8 cm; containing liver as well
1.3 Gastroschisis: evisceration through a abdominal wall defect just below and to the right, and completely separated from the umbilicus

2. Diagnosis and investigation

2.1 Antenatally by ultrasonography
2.2 CXR: Look for coexistent cardiac malformations, atelectasis, etc.

3. Transportation

Wrap the exposed bowel, or better still, the whole baby in plastic sheeting or metal foil in order to prevent severe hypothermia due to heat loss from the extruded viscera.

4. Choice of treatment

Depends on:
4.1 Birth weight/general status
4.2 Presence of associated malformations* like:
- ❏ Cardiovascular abnormalities (15–25%), especially TOF
- ❏ GI tract anomalies: atresia, volvulus
- ❏ Lower/upper midline syndrome; Trisomy D&E

5. Treatment regimes

5.1 Conservative: Paint the sac with escharotic, e.g. 0.5% mercurochrome in 65% alcohol at hourly intervals for the first 48 hours, then daily till a solid eschar forms.

Indications:
- ❏ Newborns with giant exomphalos associated with other malformations which are:
  - • Life-threatening, requiring early correction,
  - • Not compatible with much chance of survival, or
  - • Complicating the repair of exomphalos
- ❏ For salvaging failed prosthetic repair

Disadvantages:
- ❏ Prolonged hospitalization with attendant risks of cross-infection
- ❏ Mercury toxicity; safer to use instead 65% alcohol
- ❏ Results in huge ventral hernia (which frequently outgrows the peritoneal cavity).

5.2 Primary repair: For small exomphalos in full-term infants without other anomalies.

5.3 Exomphalos major or large gastroschisis in otherwise fit babies: Suture the folded edges of 2 sheets of Teflon mesh to the medial edges of both rectus muscles, thereby constructing a pouch to house temporarily herniated viscera. 2 inert silastic sheets are interposed between the bowels and the mesh. At 1–3 days' intervals, part of the viscera is gently squeezed back into the abdominal cavity and kept in position by adding a lower row of stitches until closure is completed within two weeks. Assisted ventilation may need to be maintained. Meanwhile continue gastric suction and IV alimentation.

---

* More common with exomphalos than gastroschisis.

6. Post-operative complications

The most troublesome aspect of contemporary treatment (10–25% incidence):

❑ Respiratory distress (as well as IVC compression) consequent upon overly tight repair

❑ Intestinal obstruction (either mechanical or due to prolonged peristaltic dysfunction lasting up to weeks), intestinal ischaemia or infarction, enterocutaneous fistula

❑ Suture line infection or premature separation, ventral hernia

F. CONGENITAL DIAPHRAGMATIC HERNIA

1. Definition

Herniation of abdominal viscera into the pleural cavity, mostly through a left posterolateral defect (Bochdalek).

2. Clinical suspicion

2.1 Clinical features

❑ Tachypnoea, cyanosis, costal retraction, respiratory failure

❑ "Ballooned" chest/scaphoid abdomen

❑ Absent breath sound/presence of bowel sound over affected hemithorax

❑ Displacement of apex beat

❑ X-ray:
  • "Dextrocardia" on chest X-ray
  • Gas-filled loops of bowel in hemithorax
  • Paucity of abdominal bowel gas

2.2 Differential diagnosis: all with a normal abdominal gas pattern
  Congenital cystic disease of the lung
  Eventration of diaphragm
  Staphylococcal pneumonia with pneumatocoeles
  Agenesis/hypoplasia of the lungs with elevation of the diaphragm

3. Pre-operative management

3.1 Nasogastric tube to minimize gaseous distension of the bowel.

3.2 Assisted ventilation, if needed, is administered via ET tube; bagging by mask distends the bowel and should not be performed.

3.3 Umbilical artery catheter: postductal blood gas
Right radial artery cannula: preductal blood gas

3.4 Medical regime is aimed at breaking the vicious cycle constituted by:

❑ Hypoxaemia, hypercapnia, acidosis
❑ Pulmonary artery spasm, increased muscle mass
❑ Pulmonary hypertension, persistent foetal circulation
❑ Right to left shunt

Using:

❑ Mechanical ventilation with high FiO_2 (up to 100%) and high inflating pressure to achieve a PaO_2 of 100–150 mm Hg
❑ Hyperventilation to achieve an arterial pH of 7.50–7.55, and $PaCO_2$ between 25 and 30 mm Hg
❑ Bicarbonate to correct any metabolic acidosis
❑ Plasma and dopamine (2–5 μg/kg/min up to 15–20 μg/kg/min) or dobutamine to treat circulatory failure. As dopamine may aggravate pulmonary hypertension, dobutamine is preferred.
❑ Vasodilators, e.g. "Tolazoline":
  • Indications: postductal arterial PaO_2 < 60 mm Hg FiO_2 1.0, and inspiratory pressure up to 45 cm H_2O
  • Dosage: 1–2 mg/kg/hr
  • Its action is potentiated by alkalosis/paralysis
  • Side effects: increased gastric secretion/bleeding, thrombocytopenia, oliguria and hypotension (if systolic blood pressure cannot be kept above 50 mm Hg, its use must be discontinued)
❑ For patients with persistent hypercapnia: High-frequency ventilation/oscillation at rates up to 2,400 cycles may be beneficial
❑ Prophylactic contralateral chest drain is indicated for high risk cases on the above regime.

4. Surgery

Surgery is scheduled only after having a few hours' of stabilization ($PaCO_2$ < 40; haemodynamically stable). Those who continue to deteriorate will not benefit from surgery and will not survive.

4.1 Subcostal abdominal approach is preferred.

4.2 Herniated bowels are gently reduced.

4.3 Defect is closed primary, or bridged with gortex graft if too large.

4.4 Abdominal wall is forcefully stretched to augment its capacity.

4.5 Malrotation of gut is not corrected, if asymptomatic.

5. Post-operative management

5.1 Continue assisted ventilation with hyperventilation as above. Maintain postductal PaO_2 around 80–100 mm Hg. Weaning from the ventilator should be meticulous and slow.

5.2 Repeat CXR to show lung expansion/pneumothorax/position of mediastinum.

5.3 Controversy about chest drains — 2 schools of thought:
   ❏ The ipsilateral chest drain is put to 5–7 cm H_2O suction, contralateral chest drain to 8–10 cm H_2O suction. The aim is to keep the mediastinum in the midline.
   ❏ No chest drain (especially the ipsilateral one) unless tension pneumothorax develops. This allows a slower expansion of the affected lung and has been shown to be associated with a better outcome of the patients.

5.4 Watch out for deterioration (declining PaO_2) attributable to persistent foetal circulation after a "honeymoon" period lasting a few hours to a few days.

5.5 Extracorporeal membrane oxygenation has been used with success as a last resort to provide respiratory support to patients with very severe persistent pulmonary hypertension.

References

A. Neonatal Intestinal Obstruction

1.   Nixon HH. *Neonatal abdominal emergencies*. Pro R Soc Med 1971; 64:372–74.

2.   Spitz L. *Acute Abdominal Emergencies*. In Black JA (ed): *Paediatric Emergencies*. Butterworth & Co Publishers Ltd 1979; pp. 375–89.

3.   Tan CEL, Kiely E M, Agrawal M et al. *Neonatal intestinal perforation*. J Paediatr Surg 1989; 24:888–92.

4.   Cassmann G, Korner A, Wurnig P. *Transient functional obstruction of colon in neonates: An investigation by rectal manometry and biopsy*. J Paediatr Surg 1986; 21:244–45.

5.   Ahmed H, Al-Salem K, Saleem K et al. *Congenital intrinsic duodenal obstruction: Problem in the diagnosis and management*. J Paediatr Surg 1989; 24:1247–49.

6.   Olsen NN, Luck SR, Lloyd-Smith J. *The spectrum of meconium disease in infancy*. J Paediatr Surg 1982; 17:479–81.

7.   Joseph VT, Chiang KS. *Problems and pitfalls in the management of Hirshsprung's disease*. J Paediatr Surg 1988; 23:398–402.

8.   Goto S, Ikeda K, Toyobara J. *Histochemical confirmation of the acetycholinesterase activity in rectal suction biopsy for neonates with Hirshsprung's disease*. Z Kinderchir 1984; 39:246–49.

B. Necrotizing Enterocolitis

1. Pokorny WJ, Garcia-Parts JA, Barry YN. *Necrotizing enterocolitis: Incidence, operative care and outcome.* J Paediatr Surg 1986; 21:1149–54.
2. Walsh MC, Kliegman RM. *Necrotizing enterocolitis: Treatment based on staging criteria.* Paediatr Clinic North Am 1986; 33:179–201.
3. Hill HR. *Neonatal necrotizing enterocolitis: Therapeutic decisions based on clinical staging.* Ann Surg 1978; 187:1–7.
4. Gregory JR, Compbell JR, Harrison MW. *Neonatal necrotizing enterocolitis: A ten year experience.* Am J Surg 1981; 141:562–67.
5. Cheu HW, Sukarochana, LLoyd. *Peritoneal drainage for necrotizing enterocolitis.* J Paediatr Surg 1988; 23:557–61.

C. Oesophageal Atresia

1. Ohkawa H, Ochi G, Yamazaki Y, Sawaguchi S. *Clinical experience with a sucking sump catheter in the treatment of oesophageal atresia.* J Paediatr surg 1989; 24:333–35.
2. Louhimo I, Lindahl H. *Oesophageal atresia: Primary results of 500 consecutively treated patients.* J Paediatr Surg 1983; 18:217–19.
3. Randoph JG, Altman RP, Anderson KD. *Selective surgical management based upon clinical status in infants with oesophageal atresia.* J Thoracic Cardiovas Surg 1977; 74:335–42.
4. Sillen U, Hagberg S, Rubenson A. *Management of oesophageal atresia: Rreview of 16 years' experience.* J Paediatr Surg 1988; 23:805–9.
5. Spitz L, Kiely E, Brereton. *Oesophageal atresia: Five year experience with 148 cases.* J Paediatr Surg 1987; 22:103–8.

D. Anorectal Malformations

1. Tam PKH, Chan FL, Saing H. *Direct Sagittal CT Scan: A new diagnostic approach for surgical neonates.* J Paediatr Surg 1987; 22:397–400.
2. Stephens FD, Smith ED (eds). *Ano-rectal Malformations in Children.* Chicago, Year Book Medical Publishers, 1971.
3. Kiesewetter WB, Chang JHT. *Imperorate anus: A five to thirty year follow-up perspective.* Prog Paediatr Surg 1977; 10:111–25.
4. Templeton JM, O'Neill JA. *Anorectal Malformations in Paediatric Surgery.* Chicago, Year Book Medical Publisher, 1986; pp. 1022–37.

E. Exomphalos and Gastroschisis

1. Donna A, Caniano J, Brokaw B, Ginnpease ME. *An individualized approach to the management of gastroschisis.* J Paediatr Surg 1980; 25:297–99.

2. Michalevicz D, Chaimoff CH. *The management of gastroschisis.* J Paediatr Surg 1973; 8:263–70.
3. Schuster SR. *Omphalosoele and Gastroschisis.* In *Paediatric Surgery.* 4th ed. Chicago, Year Book Medical Publishers, 1986; pp. 740–63.
4. Gierup J, Olsenk L, Sundkirst K. *Aspect of the treatment of omphalocoele and gastroschisis. Twenty years clinical experience.* Zeitschrift Fur Kinderchirurgie 1982; 35:3–6.

F. Congenital Diaphragmatic Hernia

1. Adzck NS, Vacanti JP, Lillehei CW. *Foetal diaphragmatic hernia: Ultrasound diagnosis and clinical outcome in 38 cases.* J Paediatr Surg 1989; 24:654–58.
2. Hazebroek FWJ, Tibboel D, Bos AP, Pattenier AW. *Congenital diaphragmatic hernia: Impact of pre-operative stabilization.* J Paediatr Surg 1988; 23:1139–46.
3. Langer JC, Filler RM, Bohn DJ et al. *Timing of surgery for congenital diaphragmatic hernia: Is emergency operation necessary?* J Paediatr Surg 1988; 23:731–34.
4. Ein SH, Barker G, Olley P. *The pharmacologic treatment of newborn diaphragmatic hernia — A 2-year evaluation.* J Paediatr Surg 1980; 15:384–88.
5. Zaritsky A, Chernow B. *Use of catecholamines in Pediatrics.* J Pediatr 1984; 105:341–44.
6. Wiener ES. *Congenital posterolateral diaphragmatic hernia: New dimensions in management.* Surgery 1982; 92:670–75.

21
Neonatal Sepsis

CHOW Chun-bong

1. Infants at risk

That require to be screened:
- ❏ Maternal pyrexia during labour or after delivery
- ❏ Amnionitis, foul smelling amniotic fluid or baby
- ❏ Prolonged rupture of amniotic membranes > 24 hours
- ❏ Fever, hypothermia or unstable body temperature
- ❏ Early respiratory distress
- ❏ Group B streptococcus isolated from maternal high vaginal swab

(*Note*: All at-risk infants must be closely observed for at least 24 hours.)

2. Common causative agents

Group B streptococcus
E. coli
Staphylococcus aureus
Coagulase negative staphylococcus
Pseudomonas
Klebsiella
Enterococcus
Acinetobacter

3. Clinical features

3.1 Symptoms: Usually non-specific and highly variable; high index of suspicion is required for diagnosis.

- ❑ Unstable temperature, hypo- or hyperthermia
- ❑ Lethargy, irritability
- ❑ Feed intolerance
- ❑ Vomiting, diarrhoea
- ❑ Jaundice: prolonged or resurgence
- ❑ Apnoea
- ❑ Tachypnoea
- ❑ Acidosis
- ❑ Failure to thrive
- ❑ Anaemia

3.2 Clinical examination

- ❑ General condition: lethargy, irritability, jaundice, pallor, poor peripheral circulation, dehydration
- ❑ Lesions on skin, subcutaneous tissue, scalp, joints, IV drip sites
- ❑ Chest signs: insucking, grunting, adventitious sounds
- ❑ Umbilicus: redness, tenderness, thickened umbilical cord, purulent discharge
- ❑ Otitis media
- ❑ Abdominal distension, tenderness, absent bowel sound
- ❑ Hepatosplenomegaly
- ❑ Loin tenderness, enlarged kidneys
- ❑ Limb movements: pain or limitation of movement
- ❑ Spine: imple (fistula)
- ❑ Anterior fontanelle: bulging

4. Investigations

4.1 Bacterial culture

- ❑ High vaginal swab of mother
- ❑ Gastric aspirate — for Gram stain (any bacteria seen), cell count (> 4 PML/HPF) and culture — useful within first few hours of birth
- ❑ Deep ear swab
- ❑ Umbilical swab
- ❑ Urine: suprapubic tap preferred
- ❑ Blood
- ❑ CSF if baby sick
- ❑ Endotracheal tube aspirate if applicable (also for Gram stain)

4.2 Radiological

- ❑ CXR if respiratory symptoms present
- ❑ Abdominal X-ray if intrabdominal pathology suspected

4.3 CBP and DC, platelet, ESR, C-reactive protein
4.4 Blood gas
4.5 Plasma electrolyte, urea, glucose, calcium and albumin

5. Babies requiring treatment

5.1 Early respiratory distress even without evidence of sepsis
5.2 Baby who is smelly at delivery or when there is definite amnionitis.
5.3 Preterm infants with meconium-stained amniotic fluid
5.4 Persistent temperature instability > 4 hours
5.5 WBC < 6.0 or > 20×10^9/L
5.6 Any sick baby in whom sepsis is suspected.

6. Treatment

6.1 **Choice of antibiotics**
❏ First-line: penicillin/ampicillin and aminoglycoside systemically
❏ Suspected meningitis: ampicillin and a 3rd generation cephalosporin (e.g. cefotaxime, ceftazidime)
❏ Skin or umbilical sepsis, osteomyelitis, arthritis: for anti-staphylococcal chemotherapy, replace penicillin or ampicillin with cloxacillin or cefuroxime. (Add fusidin for osteomyelitis and vancomycin for multiple resistant staphylococcal infection.)
❏ Abdominal conditions (e.g. NEC): add metronidazole.
❏ Suspected nosocomial infection: choose the appropriate antibiotic basing on the prevalence of organism in the nursery.
❏ For patients who already had and those at risk of renal impairment: avoid aminoglycoside and use instead a 2nd or 3rd generation cephalosporin or aztreonam.
❏ Change antibiotics when necessary as indicated by the sensitivity of bacteria cultured from either the baby or the mother.
❏ Specific organisms:
  • Group B streptococcus: penicillin/ampicillin
  • Staphylococcus aureus: cloxacillin
  • MRSA/CNS: vancomycin
  • Pseudomonas: ceftazidime
  • Listeria monocytogenes: ampicillin
  • Anaerobes: metronidazole

6.2 Duration of antibiotic therapy

❑ Positive cultures (of deep ear swab, gastric aspirate): minimum 1 week

❑ 10 days for GBS infection or if baby is sick.

❑ 14 days if blood culture positive

❑ Meningitis: minimum 2 weeks for G+ve organism (except Streptococcus faecalis) and 3 weeks for Gm-ve organism (including Strept faecalis).

❑ Review cultures after 48–72 hours. Consider stopping antibiotics if cultures are negative, baby looks well and mother has not been given antibiotics during delivery. If mother has been given antibiotics, continue treatment for 5 days.

6.3 Other supportive measures

❑ General: Stop oral feeding if necessary, maintain fluid and electrolyte balance and glucose homeostasis.

❑ Respiratory support: adequate oxygenation and control of acidosis

❑ Circulation: maintain blood pressure and perfusion, with colloid transfusion and pressor agents (e.g. dopamine and/or dobutamine) if necessary.

❑ Transfusion of fresh frozen plasma, fresh blood, or immunoglobulin may be of benefit.

❑ Exchange transfusion with fresh blood may be of help in overwhelming sepsis.

7. Meningitis

7.1 Common organisms

❑ GBS

❑ E. coli

❑ Staph. aureus

❑ Klebsiella-enterobacter

❑ Enterococcus

❑ Pseudomonas

❑ Proteus

❑ Listeria monocytogenes

7.2 Clinical features

❑ Early onset: fulminant process involving multiple organs in the first week of life. Usually associated with maternal factors, and acquired from infected amniotic fluid or on passage through the birth canal. Organisms: GBS, E. coli.

❑ Late onset: nosocomial infection; manifest as meningitis after first week; organisms: S. aureus, Klebsiella-enterobacter, Proteus, Acinetobacter.

7.3 Investigations
❏ Microbiological:
  - CSF:
    - Microscopy, protein and glucose content

Normal Cerebrospinal Fluid Values

	Newborn	Infant
Pressure (mm H$_2$O)	50–80	40–150
RBC (per mm^3)	< 675	0–2
WBC (per mm^3)	< 14	< 10
Protein (mg/L)^a	250–900	1 m : 700
		6 m : 400
Glucose (mmol/L)^b	2.8–4.4	2.8–4.4

a: High protein values of 2500–3000 mg/L may be found in preterms and in some term infants.

b: CSF glucose concentration varies with blood glucose which is often low in newborns.

Note: Normal CSF findings do not necessarily exclude meningitis.

    - Gram's smear and bacterial culture
    - Latex test or C.I.E. for bacterial antigen
    - Viral studies especially for ECHO and Coxsackie viruses
  - Blood culture and culture from any suspected sites
❏ Haematological: CBP/DC, platelet, ESR, C-reactive protein
❏ Biochemical:
Serum electrolyte, urea, creatinine
Blood glucose
Serum and urine osmolality
❏ Radiological:
  - Ultrasound of head: useful for diagnosis of and serial monitoring of post-meningitic ventricular dilatation; useful for detecting parenchyma infarction, haemorrhage, and ventriculitis
  - CT scan: to rule out abscess, subdural effusion or areas of thrombosis, haemorrhage or infarction

7.4 Treatment
❏ Antibiotics:
Ampicillin 200–400 mg/kg/d or penicillin 400,000 u/kg/d
plus
A 3rd generation cephalosporin (e.g.: Cefotaxime 100–150 mg/kg/d or ceftazidime 150–300 mg/kg/d)
with or without
An aminoglycoside

❏ Duration of treatment: 2–3 weeks (generally, at least 2 weeks for G+ve organism except for Strept faecalis, and at least 3 weeks for G-ve organism including Strept faecalis)
❏ Supportive measures (refer to section 6.3):
   • Nil by mouth for at least 48 hours
   • Restrict fluid: to 60% of normal requirement and monitor fluid balance carefully
   • Control of seizures: phenytoin or phenobarbitone
   • Decrease cerebral oedema: mannitol
❏ Repeat LP 48 hours later for reassessment and for CSF antibiotic bactericidal level.
❏ Blood for drug and bactericidal level.
❏ Repeat LP 48 hours after stopping antibiotics.
❏ Measure head circumference daily and cerebral ultrasound 1–2 times weekly.

8. Group B streptococcal disease

8.1 Maternal colonization
❏ 5–25% lower GI and genital tract colonization
   36–59% chronic carrier
   20–25% transient
   15% intermittent
   29% indeterminate
❏ 70% of those with positive culture in second trimester will be positive at the time of delivery.
❏ Only 8.5% of those who are culture negative in second trimester will be positive at delivery.

8.2 Neonatal colonization
❏ Vertical transmission rate 50–70%
❏ Once colonized, the organism may persist for weeks or months. 50% become negative on discharge.

8.3 Neonatal invasive disease
❏ Early onset: 1% of colonized
   • Incidence: 0.2/1000 live birth
   • Associating factors:
      – Low birth weight
      – Prolonged rupture membrane
      – Peripartum febrile episodes
❏ Late onset: not associated with any perinatal factors

❏ Clinical features:

Clinical features	Early onset (< 7 days)	Late onset (> 7 days)
Mean age at onset	20 hours	24 days
Maternal complications	Frequent (60%)	Rare
Predominant pathologies	Septicaemia/ pneumonia (68%) Meningitis (33%)	Meningitis (85%) Focal sepsis
Serotypes	I, II, III	III (90%)
Symptoms and signs	Acute onset "RDS," Apnoea, Shock	Insidious onset Fever, bulging anterior fontanelle
Mortality	Pulmonary infiltrate 55%	23%

8.4 Prevention strategies

❏ Positive culture at late third trimester:
  Intrapartum ampicillin to mother at onset of labour 2 gm IVI followed by 1 gm 4 hourly until delivery if any of the followings present:
  • Prolonged rupture of membrane
  • Intrapartum fever
  • Low birth weight
  • Previous babies suffered from GBS disease

❏ Positive culture at second trimester or positive culture at third trimester but without risk factors:
  • Obtain superficial swabs (ear canal and umbilicus) for bacterial culture and closely observe baby.
  • Perform complete "sepsis workup" and start Penicillin if the baby becomes ill or if any suspicion of deterioration in baby's condition.
  • Vaccine: a type 3 capsular polysaccharide vaccine of GBS has been shown to be protective to the baby when given at 3rd trimester to pregnant women with low blood GBS antibody level.

8.5 Treatment

❏ Penicillin 300,000 units/kg/d in 3–4 doses, or
❏ Ampicillin 200 mg/kg/d in 3–4 doses
❏ Aminoglycoside may have a synergistic effect on penicillin/ ampicillin.

9. Sticky umbilical cord

9.1 Swab for culture
9.2 Cleanse thoroughly with alcohol and apply antibiotic spray.
9.3 Any surrounding erythema or induration: obtain blood culture + CBP and give systemic antibiotic (cloxacillin).

10. Sticky eye

10.1 Obtain eye swab for bacterial culture and fluorescent stain test for chlamydia. If there is purulent discharge and gonococcal opthalmitis is suspected, the discharge should be sent for urgent Gram smear.

10.2 Early onset (< 24 hours): gonococcus, staph. aureus
Treat as gonococcal with systemic penicillin, frequent local tetracycline ointment and hourly penicillin eye drops.

10.3 Intermediate onset (2–7 days): gonococcus, staph. aureus
Swab and frequent cleansing; apply chloramphenicol eye drops and tetracycline ointment.

10.4 Late onset (> 7 days): staph. aureus, chlamydia, gonococcus
As for intermediate.
Chlamydia conjunctivitis is more likely in this group and must be investigated accordingly. The treatment for chlamydia infection is tetracycline eye ointment together with a 3-week course of systemic erythromycin.

10.5 Any periorbital cellulitis should be treated with systemic antibiotics (after blood has been obtained for bacterial culture).

10.6 Confirmed gonococcal opthalmitis should be treated with topical treatment (as above) and systemic antibiotic. In view of the prevalence of penicillin-resistant gonococcus in the community, the antibiotic of choice is a 3rd generation cephalosporin (e.g. ceftazidime).

References

1.   Gomella TL, Cunningham MD (eds). *Infectious Diseases in Neonatology: Basic Management, On-call Problems, Diseases, Drugs 88/89*. Lange Clinical Manual, Prentice-Hall International Inc, 1988.

2.   Pearse RG, Roberton NRC. *Infection in the Newborn*. In Roberton NRC (ed): *Textbook of Neonatology*. 1st ed. Churchill Livingstone, 1986.

Antimicrobial Agents for Newborn Babies

The dosage given below is listed as SINGLE dose.
For the following drugs the FREQUENCY of ADMINISTRATION is:

Preterm babies (< 37/40)	1st 7 days	q12h
	2nd week to 4th week	q 8h
	After one month	q 6h
Term babies (> 37/40)	1st 48 hours	q12h
	3rd day to 2 weeks	q 8h
	2 weeks onwards	q 6h

Drug	Single dose	Frequency	Route
Ampicillin	50 mg/kg		IV/IM/PO
Benzyl penicillin	100,000 units/kg		IV/IM
Carbenicillin	100 mg/kg		IV/IM
Cloxacillin	25 mg/kg		IV/IM/PO
Piperacillin	100 mg/kg		
Gentamicin*	2.5 mg/kg		IM/IV
Netromycin*	2.5 mg/kg		IM/IV
Tobramycin*	2.5 mg/kg		IM/IV
Amikacin*	7.5 mg/kg (loading 10 mg/kg)	q12h	IV/IM
Cefamandole#	25–50 mg/kg	q 8h	IV/IM
Cefoperazone#	25–100 mg/kg	q12h	IV/IM
Cefotaxime#	50 mg/kg	q 8h	IV/IM
Cefuroxime	25 mg/kg	q 8h	IV/IM
Ceftazidime	50 mg/kg	q 8h	IV
Ceftriazone	50 mg/kg	q12h	IV/IM
Clindamycin	10 mg/kg	q 8h	IV
Erythromycin	10 mg/kg	q 8h	IV/PO
Metronidazole	7.5 mg/kg	q 8h	IV
Septrin	Trimethroprim 5 mg/kg Sulphamethoxazole 25 mg/kg	q12h	IV/oral (NOT IM)
Vancomycin	25 mg/kg loading then 10 mg/kg	q12h	IVI
Rifampicin	5 mg/kg	q12h	IVI/PO
Chloramphenicol*	12.5–25 mg/kg in preterm babies 25–50 mg/kg in term babies and daily dose not more than 75 mg/kg		IV/IM

* Serum drug level should be checked.
# Babies receiving third generation and some second generation cephalosporins should be given Vitamin K at least twice weekly.

22
Perinatal Infection

LEUNG Chi-wai

1. Routes of transmission

1.1 Transplacental
Blood-borne (viraemia, bacteraemia, fungaemia, parasitaemia)
Effects on foetus most profound during period of organogenesis in first trimester, especially when mother has no effective humoral immune response against the infective agent

1.2 Ascending
From lower genital tract across intact or disrupted membranes

1.3 Direct inoculation
During passage through birth canal by contact, inhalation or ingestion of infected material

Note:
Transplacental and ascending = TRUE intrauterine congenital infections
Direct inoculation = perinatally acquired congenital infections

2. Aetiological agents

2.1 Viral
- ❏ RNA viruses: Rubella, Enteroviruses (Coxsackieviruses, Echoviruses), Measles, Mumps, Influenza
- ❏ DNA viruses: Cytomegalovirus (CMV)*, Herpes simplex (HSV–1, HSV–2), Varicella zoster (VZV), Epstein Barr (EBV), Hepatitis B (HBV), Human parvovirus B19, Human papillomavirus (HPV), Adenoviruses, Dengue

---

* CMV is the single commonest cause of congenital infection.

❏ Retroviruses: Human immunodeficiency viruses (HIV–1, HIV–2)

2.2 Bacterial
❏ Streptococcus agalactiae (Group B Streptococcus, GBS)
❏ Listeria monocytogenes (Listeriosis)
❏ Neisseria gonorrhoea (Gonorrhoea)
❏ Treponema pallidum (Syphilis)
❏ Mycobacterium tuberculosis (Tuberculosis)
❏ Leptospira spp. (Leptospirosis)
❏ Chlamydia trachomatis
❏ Mycoplasmas
❏ Ureaplasmas

2.3 Fungal
Candida albicans commonest

2.4 Parasitic
Toxoplasma gondii (Toxoplasmosis)
Plasmodium vivax, ovale, malariae and falciparum (Malaria)

3. Clinical features

3.1 General
❏ Most congenitally infected infants will be asymptomatic at birth. 90% CMV, 65% Rubella, 75% Toxoplasma and 50% Syphilis congenital infections are clinically inapparent at birth and features of neonatal herpes may not develop for 10–14 days after birth.

3.2 Suggestive features
❏ Growth: Low birth weight, prematurity, intrauterine growth retardation, postnatal growth failure
❏ Prolonged NNJ: haemolytic anaemia, conjugated hyperbilirubinaemia
❏ Skin rash
❏ Thrombocytopenia, hepatosplenomegaly, lymphadenopathy
❏ CNS: Microcephaly, hydrocephalus, meningoencephalitis, intracerebral calcification, cerebral atrophy, hypotonia, spasticity, seizures, mental retardation
❏ Eye: Microphthalmia, keratoconjunctivitis, glaucoma, cataract, chorioretinitis
❏ Ear: Sensorineural deafness
❏ CVS: Congenital heart lesions, myocarditis
❏ Bone: Skeletal malformations, osteitis, metaphysitis
❏ Pneumonitis
❏ Hepatitis
❏ Sepsis-like picture

❏ Non-immune hydrops foetalis

3.3 Specific features

❏ CMV:
  - Periventricular calcifications, encephalitis, sensorineural deafness
  - Chorioretinitis
  - Pneumonitis
    (*Note*: The great majority are completely asymptomatic.)

❏ Rubella:
  - Corneal opacity, cataracts, pepper-and-salt retinopathy
  - PDA, peripheral PS, VSD, myocarditis
  - Sensorineural deafness
  - Metaphysitis (radioluscent vertical "celery stalking" striations)

❏ Herpes simplex:
  - Porencephalic cyst, hydranencephaly
  - Keratoconjunctivitis, retinopathy
  - Skin vesicles or scars (absent in 20% cases)
  - Sepsis-like (respiratory failure, jaundice, shock, DIC)
  - Encephalitis (seizures)

❏ Enteroviruses:
  - Sepsis-like or aseptic meningitis

❏ Influenza:
  - Sepsis-like picture

❏ Mumps:
  - Parotitis
  - ? Endocardial fibroelastosis

❏ Hepatitis B:
  - Asymptomatic, rarely hepatitis
  - Chronic carrier or chronic hepatitis as sequelae

❏ Human papillomavirus:
  - Genital or perianal condylomata acuminata
  - Laryngeal papillomas (upper airway obstruction)

❏ HIV:
  - ? Craniofacial dysmorphism
  - Polyclonal hyperglobulinaemia
  - Lymphopenia, thrombocytopenia, reduced T4 helper cells
  - Increased incidence of opportunistic infections and lymphomas
  - Neurodevelopmental delay or degeneration
  - Lymphocytic interstitial pneumonitis

❏ Human Parvovirus B19:
  - Nonimmune hydrops foetalis in second trimester
  - Stillbirth
  - Foetal aplastic crisis

❏ EB Virus:
- Monocytosis, atypical lymphocytosis
- Proteinuria
- Metaphysitis

❏ Measles:
- Measles at delivery or within first 10 days of life

❏ Varicella zoster
- First and early 2nd trimester infection:
  - Limb hypoplasia
  - Vesicular skin lesions, skin scarring
  - Cortical atrophy, encephalitis, motor-sensory paralysis
  - Microphthalmia, cataracts, chorioretinitis
  - Muscle atrophy, club foot
  - Renal anomalies
  - Infantile zoster
- Infection within last 5 days of pregnancy to 48 hours after delivery: Severe disseminated infection in neonates (30% fatal)
- Infection 5–21 days before delivery: Mild neonatal chicken-pox

❏ Group B streptococcus:
- Early-onset syndrome (≤ 1 week of age):
  - Respiratory distress syndrome, congenital pneumonia
  - Sepsis, shock
  - Meningitis (30%)
- Late-onset syndrome (> 1 week of age):
  - Meningitis
  - Focal infections (eyes, ears, sinuses, joints, bones, skin, lungs)

❏ Listeria:
- Early-onset syndrome:
  - Sepsis
  - Congenital pneumonia
  - Skin lesions (granulomatosis infantiseptica)
  - Meconium-staining in premature neonates (chorioamnionitis)
- Late-onset syndrome: meningitis

❏ Gonorrhoea:
- Ophthalmia neonatorum (first week of life)
- Septic polyarthritis
- Meningitis
- Unexplained sepsis

❏ Syphilis:
- Nonimmune hydrops foetalis
- Hypoproteinaemic oedema (nephrotic syndrome)

- Snuffles, rhagades, "eczema oris," "saddle nose"
- Eczematoid copper-coloured maculopapular rash that may progress to bullae or desquamation ("washerwoman's skin")
- Periostitis, osteochondritis, osteitis, pseudoparalysis
- Deformed nails or paronychia
- CSF pleocytosis with elevated globulins

❏ Tuberculosis:
- Primary complex in liver (transplacental spread via umbilical vein)
- Aspiration pneumonitis (usually fulminant)

❏ Chlamydia:
- Ophthalmia neonatorum (after first week of life)
- Afebrile interstitial pneumonitis
- Otitis media
- Gastroenteritis

❏ Candida:
- Sepsis-like picture
- Congenital pneumonia
- Cutaneous lesions with skin eruptions, yellow plaques over cord and placenta
- Foreign body in mother's genital tract (IUCD or cervical cerclage suture) a common predisposing factor

❏ Toxoplasma:
- Microphthalmia, cataracts, chorioretinitis (macular lesion), optic atrophy
- Microcephaly or hydrocephalus, scattered intracerebral calcifications
- Pneumonitis
- Lymphadenopathy
- Eosinophilia
- CSF xanthochromia (high protein content) and monocytosis

❏ Malaria:
- May be asymptomatic at birth (incubation period 8–30 days)
- Fever
- Haemolytic jaundice
- Acute renal failure
- Respiratory and CNS dysfunction

4. Diagnosis

4.1 Clinical
❏ Diagnosis of the maternal infection — most useful for identifying newborns at risk but most maternal infections are asymptomatic.
❏ Maternal flu-like illness during pregnancy (e.g. CMV, Rubella,

EBV, Influenza, Enteroviruses, Human Parvovirus B19, Listeria, Toxoplasma).
❏ Recognition of a particular syndrome when clinical features are full-blown (exception rather than the rule).
❏ Ophthalmologic and audiologic evaluation when indicated.

4.2 Laboratory
❏ Non-specific tests
  • Full blood count
  • Total IgM (cord or serial postnatal blood) — a useful screening test when value raised but normal value does not exclude perinatal infection.
  • X-ray chest, skull and long bones
  • Blood culture, CSF culture
  • Ultrasound/CT brain
❏ Specific tests
  • Serology
  1. Serological results must be interpreted with caution because:
     – Maternal IgG will complicate the interpretation of organism-specific IgG results.
     – Organism-specific IgM may not be detectable in all cases (e.g. IgM for CMV or Toxoplasma may be detected in only about ½ of all congenitally infected neonates).
     – Serial serologic follow-up during the first 6 months will usually clarify the diagnosis as passively acquired maternal IgG will disappear in uninfected infants (except HIV antibody which may persist for up to 18 months). For doubtful cases serum samples should be frozen for simultaneous assay at a later stage (6–15 months) before an infection can be definitely diagnosed or excluded.
  2. Specific IgM and IgG titres for CMV, Rubella, HSV, VZV, Human Parvovirus B19, EBV (anti-VCA), Treponema pallidum,* Chlamydia,Toxoplasma gondii.**
  3. Specific IgG titre for HIV–1, HIV–2
  4. Anti-HBc IgM for HBV

---

  * RPR (rapid plasma reagin) and VDRL are non-treponemal tests for syphilis; FTA-ABS, TPHA and TPI are treponemal specific tests.
  ** Toxoplasma specific IgG tests include Sabin-Feldman dye test, ELISA, Haemagglutination and indirect immunofluorescence, a titre of > 1:25 and rising after a few weeks time is suggestive of infection.

5. Unless clinical suspicion is heightened by the occurrence of an epidemic due to a known serotype, serology for Enteroviruses is impractical.

6. CSF for VDRL in case of congenital neurosyphilis

- Antigen detection (rapid diagnosis)
  - HBsAg for HBV (sera of mother and neonate at birth)
  - DEAFF test or fluorescent antibody for CMV (saliva, urine, CSF, nasopharyngeal and conjunctival secretions)
  - ELISA or immunofluorescence for HSV (cytologic smears from cervical swab or genital lesions of mother and skin vesicles or oral ulcer/conjunctival scrapings of infant)
  - Fluorescent antibody for chlamydia (conjunctival scraping or nasopharyngeal secretion of infant, cervical swab of mother)
  - Slide coagglutination for GBS (HVS of mother, blood, CSF or concentrated urine of infant)

- DNA probes (in situ hybridization)
  - Serum, urine, respiratory secretions, foetal tissues and placenta for Human Parvovirus B19
  - Cultured lymphocytes for HIV

- Microscopy (Cytology & Histology)
  - Light microscopy for owl-eye cells in urine for CMV
  - Dark-field examination of secretion from nose, skin or mucous membrane lesions for Treponema pallidum
  - Electron microscopy for CMV in urine, HSV in scrapings from conjunctiva or skin lesions and VZV in vesicular fluid
  - Giemsa stain for inclusion bodies in conjunctival scrapings for chlamydia
  - Giemsa stain for giant cells in scrapings from vesicle for HSV (Tzanck smear)
  - Gram smear for GBS and Listeria in high vaginal swab of mother and gastric aspirate or CSF of neonate
  - Gram smear for gonococcus in conjunctival discharge
  - Ziehl-Nielsen stain for AFB in gastric aspirate and washings
  - Placental histology for candida, tuberculosis, malaria and toxoplasma
  - Liver and bone marrow biopsy for tuberculosis
  - Thick and thin peripheral blood smears for malaria

- Culture
  - CMV: Congenital CMV infection is best diagnosed in the first 2–3 weeks of life by isolating the virus in the urine (most useful), saliva, nasopharyngeal secretion or CSF
  - Rubella: Nasopharyngeal secretion, urine, stool, CSF

- HSV: Conjunctival or corneal scraping/swab, throat swab, vesicular scraping or fluid, urine, stool, CSF, maternal cervical swab or scraping from genital lesion
- HIV: Peripheral lymphocyte culture for in situ hybridization
- VZV: Vesicular fluid (late congenital chickenpox)
- Enteroviruses: Throat swab, rectal swab, stool
- GBS, Listeria: Amniotic fluid, maternal HVS, blood CSF, gastric aspirate
- Gonorrhoea: Conjunctival swab; blood, CSF
- Tuberculosis: Gastric washings, laryngeal swab
- Chlamydia: Conjunctival scraping, nasopharyngeal aspirate, genital swab, rectal swab
- Candida: Blood, urine, CSF, lower respiratory tract secretion
• Skin test
- Tuberculosis: Periodic Mantoux test at birth, 4–6 weeks, 3–4 months and 6 months of age in infant whose mother has tuberculosis

5. Treatment

5.1 GBS
❑ Penicillin G 50,000–100,000 U/kg/D iv in 4 divided doses × 10–14 days, increase to 150,000–200,000 U/kg/D iv in 4–6 divided doses if meningitis present.
❑ Ampicillin 75–100 mg/kg/D im/iv in 4–6 divided doses × 10–14 days, increase to 150–200 mg/kg/D if meningitis present.

5.2 Listeria
❑ Ampicillin 100–200 mg/kg/D iv in 4–6 divided doses *PLUS* aminoglycoside × 10–14 days (in vitro synergism)
❑ Erythromycin 40–50 mg/kg/D × 7–10 days if allergic to penicillin
❑ If meningitis present, continue for at least 1 week after resolution of symptoms

5.3 Gonorrhoea
❑ Asymptomatic infant of untreated mother: Single dose of Penicillin G 50,000 U iv/im for full-term, decrease dose for premature infants.
❑ Gonococcal ophthalmia: Penicillin G 100,000 U/kg/D iv in 4 divided doses × 7 days *PLUS* frequent eye irrigation (penicillin eye drops optional)
❑ Disseminated infection: Penicillin G 100,000 U/kg/D iv or Ampicillin 50 mg/kg/D iv/im in divided doses × 7–10 days

❑ Penicillin-resistance: Ceftriaxone 25–50 mg/kg (max. 125 mg) iv/im once or daily (or ceftazidime 150 mg/kg/day in 4 divided doses) until cured.

(*Note*: In view of the high prevalence of penicillin-resistant gonococcus in this locality, a 3rd generation cephalosporin may be the drug of first choice.)

5.4 Syphilis

❑ Evaluation required for neonates born to seropositive mothers (confirmed by treponemal test) who have:
  • Untreated syphilis
  • Treatment for syphilis within the last 30 days before delivery
  • Treatment for syphilis during pregnancy with a non-penicillin regimen
  • Absence of the expected reduction in VDRL or RPR titre after treatment
  • Absence of a well-documented history of treatment
  • Treatment for syphilis but have insufficient serological follow-up to assess disease activity.

❑ Treatment required for neonates with:
  • Active disease on physical examination or X-ray findings
  • Reactive VDRL in CSF
  • Abnormal CSF findings (WBC > 5/mm^3, protein > 0.5 g/l) regardless of CSF serology
  • Quantitative VDRL titre ≥ 4× that of the mother
  • Positive FTA-ABS IgM antibody
  • Unreliable history of duration of disease and treatment in mother

(*Note*: If in doubt, infection should be assumed and START TREATMENT! A lumbar puncture before treatment is considered mandatory.)

❑ Treatment regimen:
  • Symptomatic
    – Penicillin G 100,000–150,000 U/kg/D iv in 2–3 divided doses or Procaine Penicillin G 50,000 U/kg im daily × 10–14 days; 21 days for congenital neurosyphilis
    – If more than 1 day of therapy is missed, the entire course should be restarted (due to regeneration of spirochaetes) and inhibitory level must be maintained for at least 7 days.
  • Asymptomatic
    – Penicillin G 100,000–150,000 U/kg/D iv in 2–3 divided doses or Procaine Penicillin G 50,000 U/kg im daily × 10 days if mother untreated, inadequately treated, does not have expected serologic response to treatment, or does not have non-treponemal tests monitored during pregnancy, or laboratory evidence of congenital infection in infant.

- Benzathine Penicillin G 50,000 U/kg im once if low risk of infection, i.e. mother treated with erythromycin only during pregnancy or asymptomatic neonates with normal work-up in whom follow-up cannot be reasonably assured.
  - Erythromycin or desensitization if allergic to penicillin
❑ Follow-up:
  All treated infants should have serologic follow-up at 1, 2, 3, 6 and 12 months of age to ensure adequate response, i.e. serum VDRL or RPR titre reverted to non-detectable at 6 months or decreased by 4-fold at 3 months, CSF VDRL negative at 6 months and cell counts normal by 2 years.
❑ Criteria for retreatment:
  - If clinical signs or symptoms persist
  - If VDRL or RPR titre stable or sustained increase ≥ 4-fold
  - If initially high VDRL/RPR titre fails to decrease by 4-fold by 1 year

5.5 Tuberculosis
❑ Isoniazid 10–20 mg/kg/D po/im *PLUS* rifampicin 10–20 mg/kg/D po × 9–12 months.
❑ Add Streptomycin 20–30 mg/kg/D im for first 4 weeks *PLUS* pyrazinamide 30 mg/kg/D po for initial 2 months if drug resistance likely.
❑ If isoniazid, rifampicin and pyrazinamide are given daily for the first 2 months, treatment for 6 months may be adequate.

5.6 Chlamydia
❑ Erythromycin 30–40 mg/kg/D po in 4 divided doses × 14–21 days *PLUS* erythromycin or tetracycline or sulphacetamide ophthalmic ointment qid if conjunctivitis present.

5.7 HSV
❑ Acyclovir 10 mg/kg as 1–2 hr iv infusion q8h or Vidarabine 15–30 mg/kg/D as 12 hr or longer iv infusion × 10 days; *PLUS* trifluoridine ophthalmic solution q2h for conjunctivitis.

5.8 VZV
❑ Maternal infection in last 5 days of pregnancy to 48 hr after delivery: ZIG (zoster immune globulin) within 72 hr after birth *PLUS* Acyclovir 10 mg/kg iv infusion q8h × 10 days.
❑ Maternal infection 5–21 days before delivery: Treat varicella in infant symptomatically as for ordinary case of chickenpox.

5.9 Candida
❑ Amphotericin B 1 mg/kg/D (maximum dose) iv infusion over 4–6 hr *PLUS* 5-Fluorocytosine 150 mg/kg/D (maximum dose) iv/po in divided doses.
❑ iv Fluconazole may be a promising agent but no clinical data to support its usage in neonates.

❑ Optimal duration of anti-fungal therapy remains largely un-
known, usually 4–6 weeks required.

5.10 Toxoplasma

❑ Exact regimen for best results not yet established.

❑ Pyrimethamine 1 mg/kg/D po in 2 divided doses *PLUS* trisul-
phapyrimidines or sulphadiazine 100–150 mg/kg/D po in 4
divided doses × 30 days, folinic acid 5 mg im/po 2 times/week;
FOLLOWED BY spiramycin 100 mg/kg/D po in 2 divided
doses × 30 days in alternating courses with pyrimethamine/
sulphadiazine to minimize haematologic toxicity.

❑ May repeat therapy up to 3 times in first year of life depending on
clinical evidence of disease activity, prolonged therapy may be
required.

❑ If inflammation clinically active, corticosteroids may be considered as adjunct.

❑ Serologic follow-up of asymptomatic non-IgM antibody-positive
infants is recommended at 4–6 weeks intervals to allow treatment
or retreatment, therapy is desirable if titre is stable or rising.

❑ Regular ophthalmologic follow-up is recommended to detect
early eye manifestations.

5.11 Malaria

❑ Chloroquine-sensitive:
  • Severely ill: Chloroquine 5 mg/kg iv, then repeat in 12–24 hr
    depending on severity of symptoms.
  • Mildly ill: Chloroquine 10 mg/kg po, then 5 mg/kg po at 6, 24
    and 48 hr.
  • Follow-up with weekly chloroquine 37.5 mg (base) po
    chemoprophylaxis.

❑ Chloroquine-resistant:
  • Severely ill: Quinine 25 mg/kg/D, give 1/3 of daily dose iv
    over 2–4 hours, repeat q8h until oral therapy feasible, then
    20–30 mg/kg/D po in 3 divided doses to complete 7–10 days
    therapy.
  • Mildly ill: Quinine po as above × 7–10 days.

❑ No follow-up treatment with primaquine required as there is no
exoerythrocytic stage for congenital P.vivax and P.ovale infection.

❑ Exchange transfusion may be considered in fulminant infection.

References

1.   Sever JL, Larsen JW, Grossman JH. *Handbook of Perinatal Infections*. 2nd ed. Boston/Toronto, Little, Brown and Company, 1989.

2. Freij BJ, Sever JL (eds). *Infectious Complications of Pregnancy*. Clin Perinatol 1988; 15(2):163–363.

3. Kinney JS, Kumar ML. *Should we expand the TORCH complex?* Clin Perinatol 1988; 15:727–44.

4. Alpert G, Plotkin SA. *A practical guide to the diagnosis of congenital infections in the newborn infant*. Pediatr Clin North Am 1986; 33:465–79.

5. Nelson JD. *1991–1992 Pocketbook of Pediatric Antimicrobial Therapy*. 9th ed. Baltimore/Hong Kong/London/Sydney, Williams & Wilkins, 1991.

23

Human Immunodeficiency Virus (HIV) Infection and Acquired Immunodeficiency Syndrome (AIDS)

CHOW Chun-bong

1. Organism

HIV is a retrovirus which selectively infect T4 helper lymphocytes and macrophages.

2. Epidemiology*

3 patterns:

2.1 U.S., Western Europe, New Zealand, Australia: Prevalent among homosexuals, bisexuals, IV drug abusers; perinatal infection uncommon

2.2 South Africa, Caribbean: Heterosexual and perinatal (vertical) transmission; IV drug abuse uncommon

2.3 Asia, South-East Asia, Northern Africa, Eastern Europe: Introduced from the West — prostitutes, drug abusers

---

* Prevalence in children is increasing. It has been estimated that by 1991, most new cases of AIDS worldwide will be in the paediatric population.

3. Mode of transmission in children

3.1 According to statistics in U.S.A. in 1988:

Perinatally acquired	78%
Transfusion acquired	13%
Haemophilia	6%
Unknown	4%

3.2 Vertical transmission rate: 25–40%, higher if mothers have advanced disease.
- ❏ There is no conclusive evidence of congenital HIV syndrome exists.
- ❏ There is no clear-cut association with obstetrical complications, low birth weight, or prematurity.
- ❏ Breast feeding has been reported to transmit infection.
- ❏ Caesarean section does not prevent transmission.
- ❏ Casual contacts between mother and baby do not lead to infection.

4. Incubation period

Highly variable:
4.1 Seroconversion: 5–12 weeks, may be up to 24 months.
4.2 Perinatally infected: Median 8 months, 50% by first year, 78% by second year
4.3 Post-transfusion: Median 8–24 months

5. Clinical manifestation

In a study of 1,026 perinatally acquired AIDS cases:

Pneumocystis carinii pneumonia	34%
Lymphoid interstitial pneumonitis	28%
Recurrent bacterial infections	24%
HIV wasting syndrome	16%
Candida oesophagitis	13%
HIV encephalopathy	11%
CMV disease	7%
Pulmonary candidiasis	5%
Cryptosporidiosis	3%
Herpes simplex disease	3%
Mycobacterium avium infection	3%

(*Note*: Some had more than one disease.)

6. Diagnosis

6.1 By clinical, immunological and serological findings, and by ex-
clusion of primary immunodeficiency diseases or secondary immunodeficiency associated with immunosuppressive therapy,
lymphoreticular malignancy, or malnutrition.

6.2 Difficult in infants because of transplacental transfer of maternal
antibodies

6.3 Provisional WHO clinical case definition for childhood AIDS
- ❏ Major signs:
  - Weight loss or failure to thrive
  - Chronic diarrhoea > 1 month
  - Chronic fever > 1 month
- ❏ Minor signs:
  - Generalized lymphadenopathy
  - Oral thrush
  - Repeated common infection (otitis, pharyngitis)
  - Generalized dermatitis
- ❏ Confirmed maternal HIV infection
  (*Note*: Suspected in a child with at least 2 major and 2 minor
  signs in absence of a known cause of immunodeficiency.)

6.4 HIV tests
- ❏ Screening for anti-HIV antibody by ELIZA: Highly sensitive and
  specific
- ❏ Confirmation by Western blot or fluorescent antibody test
  (*Note*: Maternal antibody can persist for 18 months and antibody
  negative window period can last for 24 months.)
- ❏ Viral antigen or viral-DNA detection under development
- ❏ Viral culture

6.5 Immunological assessment

6.6 All infants born to HIV positive mothers should be followed and
assessed regularly. The natural history of the disease is still not
clearly defined.

7. Management

7.1 Nutritional support

7.2 Appropriate antibiotics for infections

7.3 Short courses of corticosteroid for interstitial pneumonitis

7.4 Co-trimoxazole or nebulized pentamidine prophylaxis vs pneumo-
cystis carinii lung infection

7.5 Intravenous gamma-globulin

7.6 Treatment of opportunistic infections

7.7 Anti-viral therapy with azidothymidine

7.8 Immunization with DTP, IPV, MMR, PRP-D, and Pneumococcal vaccine; avoid BCG.

8. Control measures

8.1 Universal blood and body-fluid precautions
- ❏ Hand washing after physical contact with all patients
- ❏ Gloves are recommended for handling the following body fluids and performing the following procedures:
  - Blood and blood contaminated fluids
  - Intubation
  - Endoscopy
  - Dental procedures
  - Wound irrigation
  - Phlebotomy
  - Arterial puncture
  - Vascular catheter placement
  - Tracheostomy suctioning
  - Rinsing of used instruments
  - Lumbar puncture
  - Puncture of other cavities (e.g. pleural, peritoneal)

  (*Note*: Gloves made of vinyl or copolymer do not provide adequate protection for handling blood. Surgical or latex gloves should be worn.)
- ❏ Hand washing is recommended for handling the following body fluids and performing the following procedures:
  - Urine
  - Stool
  - Vomitus
  - Tears
  - Nasal secretions
  - Oral secretions
  - Diaper change
- ❏ Prevention of needleprick injury is most important.

8.2 Childbirth
- ❏ Intrapartum care and delivery:
  Gloves, gowns over a plastic apron, boots, masks and goggles should be worn by staff assisting in childbirth and resuscitation when a woman is known to be HIV positive.
- ❏ Care of the newborn infant in labour room:
  - Gloves, gowns, mask and clear glasses for those not wearing spectacles should be worn during resuscitation.
  - Oral mucus extractors should not be used. Only mechanical suction should be used.

- Cord blood sampling: Disposable funnel may be used to reduce contamination of the sides of the collecting tubes. When needle and syringe is used, great care must be taken to avoid needle prick injury.
- Washing of newborn babies: The infant should be washed free of blood and amniotic fluid with soap and water after delivery at the labour ward. Care must be taken to avoid hypothermia.

❏ Postnatal care of newborn baby:

Once the baby of an HIV positive or high risk mother is washed free of blood, he can be handled in the normal way without gloves, and should be cared for in the same room as his mother. The universal blood and body-fluid precautions, however, should still be applied.

8.3 Exposed health care workers

❏ Risk of infection:

- Needleprick injury: rate of infection estimated to be 0.4%.
- Mucous membrane exposure: no increased risk.
- Inoculation of skin lesion or contamination of intact skin: no increased risk has been detected so far in any of the prospective studies.

❏ After exposure, the followings should be performed:

- Confirm that the patient is HIV-positive.
- Evaluate clinically and serologically for HIV infection at 0, 3, 6, 12 months.
- Counselling should be provided.

References

1. Oxtoby MJ. *Perinatally acquired human immunodeficiency virus infection.* Pediatr Infect Dis J 1990; 9:609–19.
2. *Update: Universal precautions for prevention of transmission of HIV, HBV and other bloodborne pathogens in health-care setting.* MMWR 1988; 38:3S–18S.
3. Becker CE. *Occupational infection with HIV: Risks and risk reduction.* Annuals Int Med 1989; 110:653.
4. Lissauer T. *Impact of AIDS on neonatal care.* Arch Dis Child 1989; 64:4–7.
5. Nicholas SW, Sondheimer DL, Willoughby AD et al. *Human immunodeficiency virus infection in childhood, adolescence, and pregnancy: A status report and national research agenda.* Pediatrics 1989; 83:193–308.

24

Neonatal Haematology

LI Shan-ho

1. Normal values

1.1 Haemoglobin

| | Haemoglobin (g/dL) | |
Age	Term	Preterm
0	15–17	14.5

1.2 White cell count

| | Neutrophil ($\times 10^9$/L) | | Lymphocyte ($\times 10^9$/L) | |
Age	Term	Preterm	Term	Preterm
0	11 (5–26)	5 (2–9)	5.5(2–11)	4 (2.5–6)
24 h	11 (5–21)	7.5(3–9)	5 (2–9)	3.5(1.5–3)
72 h	5.5(2–7)	4.5(3–7)	3.5(2–5)	3 (1.5–4)
1 wk	5 (2–8)	3.5(2–7)	5 (3–6)	4.3(2.5–7)
1 mo	3.8(1–9)	2.5(1–9)	6 (3–15)	6.5(2–15)

2. Anaemia

2.1 Definition
Central venous haemoglobin < 13 g/dL or capillary venous haemoglobin < 14.5 g/dL.

2.2 Causes
❑ Haemorrhagic anaemia:

- Antepartum period:
  - Loss of placental integrity: Abruptio placenta, placenta previa.
  - Anomalies of umbilical cord or placental vessels: Velamentous insertion of umbilical cord, vasa previa.
  - Twin-twin transfusion.
- Intrapartum period:
  - Fetal-maternal haemorrhage.
  - Caesarean delivery.
  - Traumatic rupture of umbilical cord.
  - Obstetric trauma.
- Neonatal period:
  - Concealed haemorrhage: Caput succedaneum, cephal-haematoma, intracranial haemorrhage, visceral parenchymal haemorrhage.
  - Defects in haemostasis.
  - Repeated blood sampling.
❏ Haemolytic anaemia:
- Immune haemolysis:
  - Isoimmune: ABO, Rh or minor blood groups (c, E, Kell) incompatibility.
  - Autoimmune.
- Nonimmune haemolysis:
  - Bacterial sepsis may cause primary microangiopathic haemolysis.
  - Congenital viral infections (e.g. TORCH).
- Congenital erythrocyte defect:
  - Metabolic enzyme deficiency: G6PD, pyruvate kinase.
  - Haemoglobinopathy.
  - Membrane defects: Hereditary spherocytosis.
❏ Hypoplastic anaemia

2.3 Clinical assessment

❏ Anaemia at birth:
- Haemorrhagic anaemia:
  - Unusual blood loss in labour: Examine placenta for abnormal vessel.
  - Signs of concealed bleeding.
  - Signs of twin-twin transfusion: Compare haematocrits.
  - Signs of shock: Rising pulse rate, tachypnoea.
  - Always measure baby's blood pressure.
- Haemolytic anaemia: IUGR, hydrops fetalis and Rh-negative mother.
❏ Anaemia presenting after 24 hours of age:
Often associated with obstetric trauma, poor placental transfusion, and unrecognized perinatal haemorrhage.

❑ Anaemia associated with jaundice:
Suggests haemolytic anaemia.

2.4 Laboratory investigation

❑ Initial studies:
- Haemoglobin/Haematocrit: Takes time to fall after haemorrhage.
- Reticulocyte count: Elevated in antecedent chronic haemorrhage or haemolytic anaemia.
- Blood smear.
- Direct Coombs test: Positive in isoimmune or autoimmune haemolysis.

❑ Selected laboratory studies:
- Isoimmune haemolysis: ABO and Rh blood group of both baby and mother/parents.
- Foetal-maternal haemorrhage: Kleihauer–Betke test.
- Coagulation studies.
- Screening for TORCH and other viral infections.

2.5 Management

❑ Simple replacement transfusion — indications:
- Acute haemorrhagic anaemia.
- Ongoing deficit replacement.
- Maintenance of effective oxygen-carrying capacity in:
  - High-risk infants with apnoea or compromised tissue oxygenation (keep Hb > 15 g/dL).
  - Healthy low-birth weight infants, keep Hb > 12 g/d L. (Owing to the potential hazards of blood transfusion, the Hb level at which transfusion is indicated remains a controversy.)

❑ Emergency transfusion at birth:
- Use group O, Rh-negative packed red blood cells.
- Alternative replacement fluids include fresh-frozen plasma, 5% albumin, and dextran.
- Catheterize umbilical vein.
- Take blood samples for diagnostic studies.
- Infuse 10–15 ml/kg of replacement fluid over 10–15 minutes.
- If simple transfusion is indicated, calculate the volume of packed red blood cells needed by the following formula:

$$\frac{\text{weight [kg]} \times 80 \text{ ml/kg} \times (\text{desired Hct} - \text{patient's Hct})}{\text{Hct of donor packed red cells}}$$

- Monitor central venous pressure when a single transfusion exceed 10 ml/kg.

❑ Exchange transfusion — indications:
- Chronic haemolytic anaemia.

- Haemorrhagic anaemia with raised central venous pressure.
- Severe isoimmune haemolytic anaemia with circulating sensitized red blood cells and antibody.
- Disseminated intravascular coagulation (DIC).

❏ Nutritional replacement — indications:
- Iron:
  - Foetal-maternal haemorrhage of significant volume.
  - Chronic twin-twin transfusion (donor twin).
- Folate:
  - Preterm infants weighing less than 1.5 kg.
  - Chronic haemolytic anaemia.
  - Infants receiving phenytoin.
- Vitamin E: Infants below 1.5 kg.

❏ Prophylactic nutritional supplementation
- Elemental iron: 1–2 mg/kg beginning at 2 months of age and continuing through 1 year of age.
- Folic acid: 1–2 mg/day for preterm infants; 5 mg/wk for term infants.
- Vitamin E: 25 iu/day until a corrected age of 4 months.

3. Coagulation disorder

3.1 Normal values

Test	Adult and older child	Term infant	Preterm infant
Platelet count ($\times 10^9$/L)	150–400	150–400	150–400
Prothrombin time (sec)	10–14	11–15	12–16
Partial thromboplastin time (sec)	25–35	30–40	30–80
Thrombin time (sec)	15–20	15–20	17–25
Fibrinogen (mg/dL)	175–400	175–350	150–325
FDP (μg/ml)	< 10	< 10	< 10

3.2 Causes

❏ Deficiency of vitamin K-dependent coagulation factors (factor II, VII, IX, X):
- No vitamin K prophylaxis at birth: Typically presents on 2–4 days of age.
- Prolonged breast feeding.
- Use of antibiotics.

❏ Consumptive coagulopathy (DIC):

- Disseminated congenital or viral infection.
- Bacterial sepsis.
- Prolonged hypotension/acidaemia.
- Hypothermia.
- Severe birth asphyxia.

❏ Congenital factor deficiency: Haemophilia (factor VIII deficiency) accounts for 90% of such cases.

❏ Thrombocytopenia:
- DIC.
- Maternal idiopathic thrombocytopenia or SLE.
- Congenital infection.
- Isoimmune thrombocytopenia.
- Maternal drugs (quinidine, thiazide).

3.3 Diagnostic approach
(Refer to Fig. 1.)

3.4 Management
❏ Haemorrhagic disease of newborn:
- Immediate replacement transfusion if in shock.
- Fresh frozen plasma.
- Vitamin K_1 1 mg intravenously.

❏ Disseminated intravascular coagulation:
- Treat underlying disease (correct hypotension or acidaemia).
- Replace clotting factors by fresh frozen plasma transfusion (10–15 ml/kg).
- In severely affected infants, exchange transfusion with fresh blood (not more than 48 hours after collection).

❏ Congenital factor deficiency:
Replace missing factor by transfusing factor concentrate, cryoprecipitate.

❏ Thrombocytopenia:
- Obstetric management of maternal autoimmune thrombocytopenia:
  – Aims at preventing intracranial haemorrhage during vaginal delivery.
  – Maternal corticosteroid and intravenous gamma-globulin have been used for raising foetal platelet count.
  – Antenatal cordocentesis and foetal scalp blood sampling may be employed to determine foetal platelet count.
  – Maternal antiplatelet antibody level (but not maternal platelet count) appears to correlate with severity of foetal thrombocytopenia.
  – Caesarean section for high risk cases (e.g. foetal platelet $< 50 \times 10^9$/L).
- Treatment of infants with thrombocytopenia:
  – Treat the underlying causes (e.g. sepsis).

Figure 1. Diagnostic Approach to Coagulation Disorders in Neonate

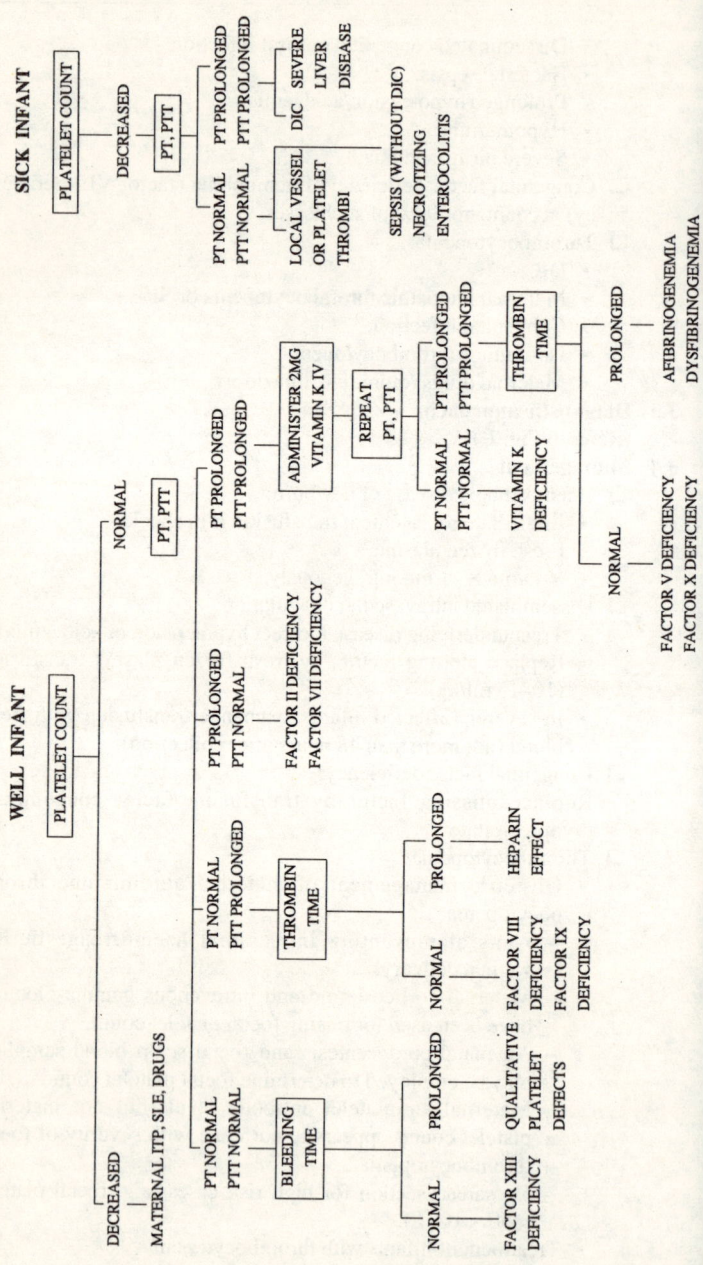

      – Intravenous gammaglobulin and corticosteroid (prednisone 2 mg/kg/day) may be used for immune thrombocytopenia.
      – Platelet transfusion for active bleeding or platelet count $< 20 \times 10^9$/L. Random donor platelets are given in a dosage of 1 unit/5 kg.
      – Exchange transfusion for immune thrombocytopenia.

4. Neonatal polycythaemia

4.1   **Definition:** Venous haematocrit $\geq$ to 65%.

4.2   **Risk factors**
- ❑ Infants of diabetic mothers.
- ❑ Intrauterine growth retardation.
- ❑ Perinatal asphyxia.
- ❑ Recipient of twin-to-twin transfusion.
- ❑ Large placental transfusion.
- ❑ Rarities: e.g. Beckwith syndrome.

4.3   **Presentation**
- ❑ CNS depression, fits and cortical venous thrombosis.
- ❑ Heart failure.
- ❑ Respiratory distress, cyanosis and persistent foetal circulation
- ❑ Jaundice, NEC.
- ❑ Hypoglycaemia; hypocalcaemia.
- ❑ Renal vein thrombosis.

4.4   **Management**
- ❑ Asymptomatic: Expectant observation; partial exchange transfusion with plasma if venous haematocrit > 70%.
- ❑ Symptomatic: Partial exchange transfusion with plasma.
- ❑ Formula for calculating blood volume to be exchanged:

$$= \frac{\text{weight [kg]} \times 80 \text{ ml/kg} \times (\text{observed Hct} - \text{desired Hct [55\%]})}{\text{observed Hct}}$$

References

1.   Nathan DG, Oski FA (eds). *Hematology of Infancy and Childhood.* 3rd ed. Philadelphia, WB Saunders, 1987.
2.   Roberton NRC (ed). *Textbook of Neonatology.* 1st ed. London, Churchill Livingstone, 1986.
3.   Gomella TL, Cunningham MD (eds). *Neonatal Basic Management, On-call Problems, Diseases, Drugs.* 1st ed. Connecticut, Appleton and Lange, 1988.

4. Roberton NRC. *A Manual of Neonatal Intensive Care*. 2nd ed. London, Edward Arnold Ltd, 1986.
5. Oh W. *Neonatal polycythemia and hyperviscosity*. Pediatr Clin North Am 1986; 33(3):523–32.
6. Buchanan GB. *Coagulation disorders in the neonate*. Pediatr Clin North Am 1986; 33(1):203–20.

25

Neonatal Jaundice, Phototherapy, Exchange Transfusion

LEE Wai-hong

A. NEONATAL JAUNDICE

1. Physiological vs pathological

Features suggesting pathological jaundice:
- Jaundice in first 48 hours of life; or of late onset after day 5.
- Jaundice in all ill infant.
- Total serum bilirubin rising by more than 5 mg/dl (85 µmol/L) per day.
- Recurrence after initial improvement.
- Clinical jaundice persisting over 14 days.
- Direct bilirubin > 1.5 mg/dl (25 µmol/L) or < 30% of total bilirubin.

Under these circumstances, jaundice must be investigated.

Notes:
(1) "Physiological" jaundice can reach pathological level.
(2) Bilirubin levels within "physiological" range does not preclude the presence of a pathological process.

2. Pattern of neonatal jaundice

- Early onset (< 48 hours):
  Haemolytic disease of newborn
  Intrauterine infection
- Mild jaundice (day 3–5):
  Usually physiologic "breast-feeding jaundice"

Figure 1.

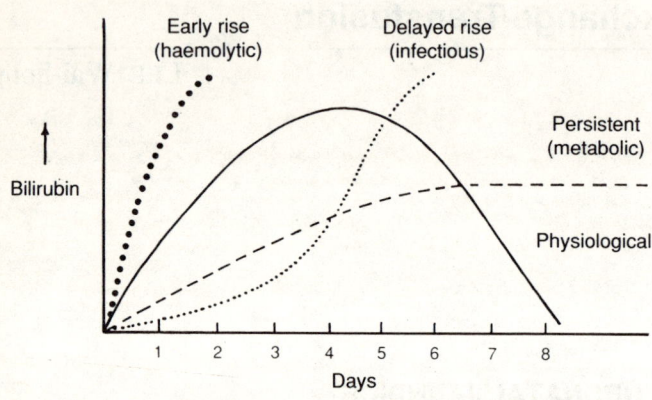

- Rapid rise (> 48 hours):
  HDN/congenital infection haemolysis in G6PD deficiency
- Persistent jaundice (> 2 weeks):
  Persistence of acute neonatal cause
  Low grade infection, UTI
  Breast milk jaundice
  Metabolic causes
  Obstructive jaundice

3. Clinical features of bilirubin encephalopathy (Kernicterus)

- Stage 1:
  Lethargy, poor sucking, vomiting
  Hypotonia, sluggish moro reflex
- Stage 2:
  Fever, irritability, high-pitched cry
  Frowning, sunsetting eyes, oculomotor paresis
  Hypertonia, opisthotonus, seizures
  Absent moro reflex, cardio-respiratory arrest
- Stage 3:
  Jaundice subsides, CNS signs disappears
- Stage 4:
  (Late Sequelae)

Choreo-athetosis/rigidity
High tone deafness (sensorineural)
Speech disorders
Mental retardation
Squints, Visual-motor incoordination
Greenish staining of teeth

4. Work-up of non-physiologic jaundice

- ❏ Serum bilirubin (direct & indirect)
- ❏ Urinalysis
  - Bilirubin indicates obstructive jaundice
  - Increased urobilinogen not usually seen in haemolysis
- ❏ Stool for colour
- ❏ Hb & retic count
- ❏ Blood smear for RBC morphology
- ❏ Blood group & Rh typing if HDN suspected
- ❏ Coombs' test (may be negative in ABO & in anti-M)
- ❏ Identification of antibody from maternal sera & infant RBC elvate, indicated if
  - Positive Coombs' test
  - Early onset jaundice within 48 hours
  - Strong clinical evidence of HDN
- ❏ Urine for reducing substance
- ❏ Viral culture & serology
- ❏ Bacterial sepsis screen (including urine culture)
- ❏ Liver function & thyroid function if jaundice prolonged

5. Work-up for conjugated hyperbilirubinaemia

- ❏ Evaluate liver function for impairment: Liver enzymes, albumin, prothrombin time, bile acids
- ❏ Stool colour & stercobilinogen
- ❏ Urinalysis for bilirubin
- ❏ Sepsis screen including urine & blood cultures
- ❏ Virology screen for TORCHES
- ❏ X-ray skull & long bones (as indicated)
- ❏ Urine metabolic screen for reducing substances & amino acids
- ❏ Serum α_1 antitrypsin
- ❏ T3/T4, TSH, glucose
- ❏ Sweat chloride (as indicated)
- ❏ Abdominal ultrasonography for structural lesions
- ❏ Radionuclide hepatobiliary imaging: Locally Rose Bengal not

available; usually an iminodiacetic acid derivative used, e.g.
EHIDA
❏ Duodenal intubation (colour, bilirubin, bile acids): May
demonstrate complete obstruction.
❏ Percutaneous liver biopsy
Note: Aim to achieve working diagnosis before 6 weeks in order to
avoid delay in timing of Kasai Operation.

6. Management

6.1 General
❏ Ensure adequate intake and hydration
❏ Discontinue any medication which might interfere with bilirubin
metabolism or binding
❏ Correct any factor that might make CNS more susceptible to
bilirubin toxicity
❏ Closely monitor bilirubin level and clinical state
❏ Use phototherapy to increase bilirubin excretion when bilirubin
level may become hazardous were it to rise further
❏ Use exchange transfusion to remove bilirubin when potentially
toxic levels are reached
❏ Define aetiology
6.2 Follow-up care after severe neonatal jaundice
❏ Check for late anaemia (usually 2–4 weeks)
❏ Screen for hearing (auditory screening cradle, or BAEP)
❏ Developmental follow-up

Table 1. Diagnostic Approach to Neonatal Jaundice

	Data	Significance
Family history	Jaundice, anaemia, early gall bladder disease, splenectomy	Hereditary haemolytic anaemia such as spherocytosis
	Sibling with NNJ & anaemia	Suggests blood group incompatibility, G6PD deficiency
	Sibling with NNJ while breast feeding	Suggests breast milk jaundice
	Liver disease in family	Suggests metabolic causes of NNJ, e.g. α_1 antitrypsin deficiency, galactosaemia etc.
Maternal history	Febrile illness or rashes during pregnancy	Suggests intrauterine infections
	Previous transfusions, pregnancies & abortions	May initiate maternal isoimmunization in HDN
	Maternal drugs like sulphonamides, nitrofurantoin, antimalarials	May cause haemolysis in G6PD deficient baby
	Gestational DM	IDM prone to NNJ
Labour & delivery	Traumatic delivery	Suggests intracranial haemorrhages, sequestered blood
	Oxytocin	Associated with NNJ
	Delayed cord clamping	Leads to polycythaemia
Baby's history	Small for gestational age	Polycythaemia Intrauterine infections
	Delayed passage of meconium	Increases enterohepatic circulation of bilirubin

Table 1. Diagnostic Approach to Neonatal Jaundice (cont'd)

	Data	Significance
Baby's history (cont'd)	Constipation	Consider intestinal obstruction, hirchsprung, meconium plug
	Persistent vomiting	Consider sepsis, pyloric stenosis, intestinal obstruction, galactosaemia
	Poor caloric intake	Decreases bowel motility Delays hepatic conjugation Dehydration leads to haemoconcentration & increased bilirubin level
	Breast feeding	First week: "breast feeding" jaundice due to inadequate intake Late: suggests true breast milk jaundice
Clinical signs in baby	Signs of IUGR	Consider intrauterine infections, polycythaemia
	Trisomy features	Hepatitis with Trisomy 13 or 18
	Small head	Suggests intrauterine infection
	Facies	May suggest cretin, Alagille
	Cephalhaematoma, subaponeurotic haematoma Bruising of body	Sequestered blood as cause of NNJ
	Pallor	Haemolytic disease, or sequestered blood

Table 1. Diagnostic Approach to Neonatal Jaundice (cont'd)

	Data	Significance
Clinical signs in baby (cont'd)	Petechiae	Intrauterine infection, sepsis, erythroblastosis
	Plethora	Polycythaemia contributes to NNJ (especially IUGR, infant of diabetic mother)
	Hepatosplenomegaly	Suspect haemolytic anaemia, intrauterine infection, postnatal sepsis, liver disease
	Mass below liver	Choledochal cyst
	Umbilical hernia	Consider hypothyroidism
	Yellow discoloration of abdominal wall, abdominal distension, hernia/hydrocoeles	Suspect spontaneous perforation of bile duct
	Micropenis	Suspect cholestasis & hepatitis associated with hypopituitarism
	Cataracts	Congenital infection, galactosaemia
	Chorio-retinitis	Congenital infection
	Dark urine, clay coloured stools	Obstructive jaundice

B. PHOTOTHERAPY

1. Guide to use of phototherapy

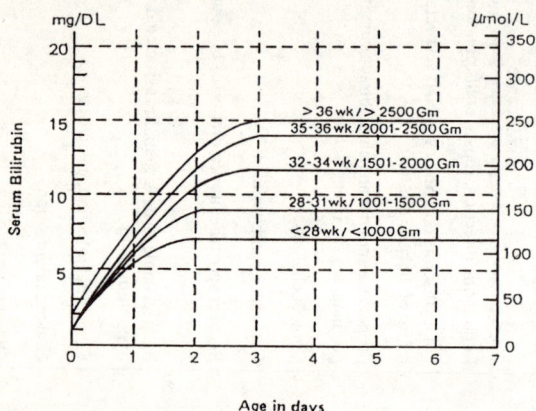

Age in days

2. Methodology

- ❑ White, blue, green lights have been used
- ❑ Most effective wavelength: 425–475 nm
- ❑ Effective Spectral Irradiance: 6–12 $\mu w/cm^2/nm$ in 425–475 nm range
- ❑ Light source positioned 50 cm above infant
- ❑ Unidirectional vs bidirectional
- ❑ Continuous vs intermittent
- ❑ Quality assurance:
  - • Check irradiance of lights periodically.
  - • Change bulbs after specified period of use as recommended by manufacturer.

3. Care of babies on phototherapy

3.1   Expose as much surface area as possible.
3.2   Shield baby's eyes from light source.
      Watch out for occlusion of nostrils.
      Check eyes for conjunctivitis/discharge.
      Remove eye shields periodically, e.g. while parent visits, during feeding or bathing.

3.3 Monitor baby temperature every 2 hours.
Watch for hyperthermia if nursed in incubator.

3.4 Monitor intake and output accurately.
Loose green stools may develop.
Check body weight at least daily.

3.5 Provide for increased fluid intake to compensate for increased insensible water loss.
❏ Extra 10–20% for term babies
❏ Additional 20–40 ml/kg/day for premies

3.6 Measure serum bilirubin at least 12-hourly, and more frequently when it is rising rapidly.

3.7 Monitor for rebound after discontinuing phototherapy.

3.8 Erythematous skin rashes may develop, but will subside when phototherapy is discontinued.

3.9 If bronzing of skin appears, check for unrecognized conjugated hyperbilirubinaemia.

C. EXCHANGE TRANSFUSION

1. Objectives

❏ To remove bilirubin
❏ To remove antibodies and sensitized RBC
❏ To correct anaemia

2. Indications

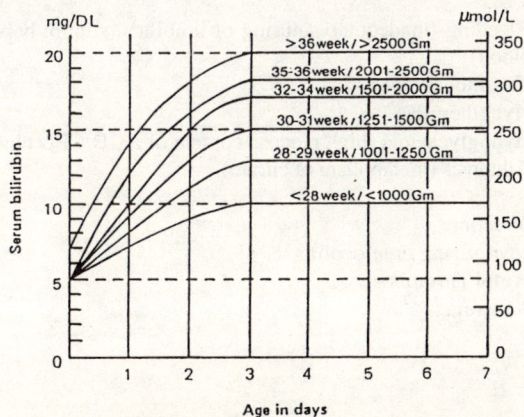

2.1 Bilirubin ≥ 20 mg/dl in term babies > 72 hours old

2.2 Term babies < 72 hours according to modified Allen-Diamond graph

2.3 Premies

According to modified Allen-Diamond graph

Alternatively, for preterm ≤ 1800 gm:

$$\text{Bilirubin level (mg \%) for exchange} = \frac{\text{Birth weight (in gm)}}{100}$$

2.4 In Rh-haemolytic disease
- ❏ Cord blood bilirubin ≥ 5 mg/dl (85 μmol/L)
- ❏ Cord blood haemoglobin < 10 g/dl
- ❏ Bilirubin rise exceeding 0.5 mg/dl per hour on day 1

2.5 Reduce level for exchange by 2–3 mg/dl in presence of risk factors, e.g. pH < 7.15, prolonged hypoxaemia, severe sepsis, signs of clinical or CNS deterioration.

3. Dangers and complications

3.1 During E.T.
- ❏ Volume imbalance (Hypovolaemia/Hypervolaemia) (technical error)
- ❏ Arrhythmias (hyperkalaemia; hypocalcaemia; catheter in atrium)
- ❏ Hypocalcaemia (citrate toxicity)
- ❏ Hyperkalaemia (old blood; overheated blood)
- ❏ Hypernatraemia
- ❏ Acidosis (ACD blood)
- ❏ Air Embolism
- ❏ Hypothermia (over exposure)
- ❏ Hypoglycaemia (heparinized blood)

3.2 After E.T.
- ❏ Bleeding (inadequate suturing of umbilical stump; heparinized blood)
- ❏ Thrombocytoperia
- ❏ Hypothermia
- ❏ Hypoglycaemia (high glucose content of ACD/CPD blood)
- ❏ Alkalosis (metabolism of citrate)

3.3 Late
- ❏ Infection
- ❏ Necrotizing Enterocolitis
- ❏ Portal Thrombosis
- ❏ Anaemia

4. Preparation for exchange: blood

4.1 Use fresh blood < 72 hours old.

4.2 Preferably measure Na/K of donor blood (blood in pilot tubing of pack can be used). Advisable to reject pack if K > 9 mmol/L, or Na > 170 mmol/L.

4.3 Warm blood in water bath at between 27–37°C. Never warm under hot tap.

4.4 Total volume exchanged should be 160–180 ml/kg (two-volume exchange).

5. Preparation of baby

5.1 Allow feeding up to shortly before exchange. Empty stomach immediately before procedure.

5.2 Connect baby to cardiorespiratory monitor. Check basal vital signs, especially body temperature.

5.3 Keep baby warm in warming cradle or preferably a servo-controlled radiant heater. Do not hang blood pack directly in the heat path of radiant warmer.

5.4 Correct hypoxia, hypoglycaemia, hypothermia and acidosis in sick infant before performing exchange.

6. Procedure

6.1 Use full aseptic precautions.

6.2 Prepare
   ❏ Heparinized normal saline (5 unit/ml) for rinsing
   ❏ 10% calcium gluconate solution
   ❏ 8.4% $NaHCO_3$ solution for ill baby

6.3 Carefully clean umbilical stump and peri-umbilical area with iodine and alcohol. Drape around umbilicus.

6.4 Umbilical vein catheterization
   ❏ Use 5FG/8FG end-hole catheter filled with heparinized saline.
   ❏ Remove any blood clots in vessel lumen before introducing catheter.
   ❏ Advance catheter until a free flow of blood is obtained. DON'T exceed 2/3 shoulder-umbilical distance to avoid entry into heart.
   ❏ Never allow catheter to open to air.

6.5   Set up a closed system as shown:

Figure 2. Set up for E.T.

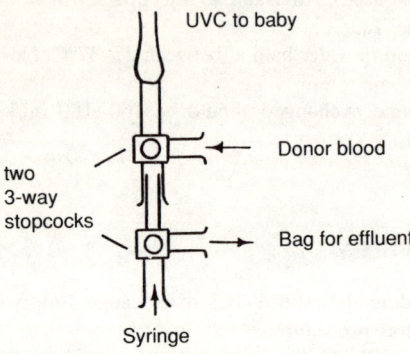

6.6   Start exchange with each cycle following the steps belows:
- ❏ Draw blood from baby SLOWLY.
- ❏ Push blood into wastage bag.
- ❏ Draw blood from donor pack.
- ❏ SLOWLY inject donor blood into umbilical vein over 1–2 min watching out for air bubbles.
- ❏ Wait for ½ min and resume cycle.
  Notes:
  (1) Each cycle should take at least 3 min.
  (2) Aliquot volume should not exceed 10% of estimated blood volume.

  $$> 2500 \text{ gm: } 20 \text{ ml}$$
  $$1801–2500 \text{ gm: } 15 \text{ ml}$$
  $$1201–1800 \text{ gm: } 10 \text{ ml}$$
  $$< 1200 \text{ gm: } \;\; 5 \text{ ml}$$

  (3) Amount of blood withdrawn or injected in each cycle must be exact.
  (4) Watch out for stopcock direction to prevent error in direction of blood flow.

6.7   Save the first and last aliquot from baby.

6.8   For every 100 ml blood exchanged, add 1 ml of 10% calcium gluconate to exchange aliquot. Inject SLOWLY.

6.9   Gently shake blood pack periodically to prevent settling of red cells.

6.10 Monitor baby's condition closely throughout. Accurately record time and volumes of blood exchanged, drugs given, and vital signs of baby.

6.11 End of exchange
- ❑ Close vein by purse-string suture of catgut and remove catheter.
- ❑ Spray stump with polybactrim and apply a light dressing.
- ❑ Unless a campelling reason exists, the venous catheter should not be left in situ.

6.12 The recommended time for exchange transfusion is between 60–90 minutes. Use slower rate for the smaller and sicker infants.

7. Specimens to be collected

7.1 Pre-E.T. umbilical swab for bacteriology.

7.2 Pre-E.T. blood for
- ❑ Hgb PCV smear reticulocyte count
- ❑ Liver function, including direct and indirect bilirubin
- ❑ Coomb's test (if not already checked)
- ❑ Blood culture
- ❑ Others (e.g. glucose in Rh incompatibility, or G6PD if not already screened).

7.3 Donor blood for

Hgb, PCV ⎫
Na, K   ⎬ as indicated
pH   ⎭

7.4 Post-E.T. blood for
- ❑ Hgb/PCV, platelets
- ❑ Bilirubin (direct and indirect)
- ❑ Blood culture
- ❑ Ca, Na, K
- ❑ Glucose
- ❑ Reserve sample for cross match.

7.5 Catheter tip at end for bacteriology.

8. Danger signs during E.T.

- ❑ Cyanosis/pallor
- ❑ Bradycardia/tachycardia
- ❑ Arrhythmia
- ❑ Grunting respiration/tachypnoea/depressed respiration
- ❑ Irritability/jitteriness
- ❑ Blood withdrawn from baby getting dark-coloured.

If any untoward signs appear, attend to baby immediately: Interrupt procedure until baby recovered.

9. Post-exchange management

- ❏ Check body temperature.
- ❏ Hourly AR and RR for 6 hours, or till steady.
- ❏ Watch for umbilical bleeding.
- ❏ Watch for rebound hypoglycaemia: check dextrostix at 2 and 4 hours after exchange.
- ❏ Resume milk feeding not later than 3–4 hours after exchange.
- ❏ Continue phototherapy.
- ❏ Recheck bilirubin 4–6 hours after exchange.
- ❏ Antibiotics need not be given as a routine after exchange transfusion, unless the umbilical area was dirty or there was some break in sterile technique.

10. Modification of technique — albumin administration

10.1 In severe hyperbilirubinaemia:
- ❏ To increase the amount of bilirubin removed by exchange transfusion to about 30–40%
- ❏ To increase the plasma albumin binding capacity

10.2 Method
- ❏ Albumin priming:
  - • 1 gm/kg of salt-poor albumin intravenously
  - • 1 hour before exchange (4 ml/kg 25% albumin)
- ❏ Albumin substitution:
  - • 6.25 gm albumin added to each 500 ml donor blood before exchange alternatively, the calculated amount of albumin may be added in divided doses to each cycle independently.

10.3 Precautions
- ❏ May precipitate heart failure.
- ❏ Contraindicated in severe anaemia and in hydropic baby.
- ❏ Do not withhold a decision for exchange if a fall in bilirubin level is seen after albumin priming.

11. Modification of technique — alternative routes

To be used when umbilical vein no longer accessible.

11.1 Push-pull technique
- ❏ Large peripheral vein (saphenous, brachial)
- ❏ Umbilical artery (if catheter already in situ)

11.2 Isovolumetric technique
Continuous removal of blood from an artery balanced by continuous
infusion into a vein:
- ❏ Peripheral artery — peripheral vein
- ❏ Umbilical artery — peripheral vein (where UAC already in situ)

References

1. Cockington RA. *A guide to the use of phototherapy in the management of neonatal hyperbilirubinaemia.* J Pediatr 1979; 95:281.
2. Committee on Phototherapy in Newborn. *Final Report.* Washington DC, National Research Council, 1974.
3. Harper RG et al. *Kernicterus — 1980.* Clin Perinat 1980; 7:75.
4. Lee KS et al. *Unconjugated hyperbilirubinaemia in very low birth weight infants.* Clin Perinat 1977; 4:305.
5. Thompson TR. *Intensive Care of Newborn Infants.* Minneapolis, University of Minnesota Press, 1982; pp. 157–64.
6. Bowman JM. *Haemolytic Disease of the Newborn.* In Roberton NRC (ed): *Textbook of Neonatology.* Edinburgh, Churchill Livingstone, 1986; pp. 469–83.
7. Cloherty JP. *Neonatal Hyperbilirubinaemia.* In Cloherty JP, Stark AR (eds): *Manual of Neonatal Care.* 2nd ed. Boston, Little, Brown, 1985; pp. 233–62.
8. Davies PA et al. *Medical Care of Newborn Babies.* London, Heinemann, 1974; pp. 274–77.
9. Haber BA, Lake AM. *Cholestatic jaundice in the newborn.* Clin Perinat 1990; 17:483–500.
10. Maisels MJ. *Neonatal Jaundice.* In Avery GB (ed): *Neonatology.* 3rd ed. Philadelphia, Lippincott, 1987; pp. 534–629.
11. Mowat AP. *Disorder of the Liver and Biliary System.* In Roberton NRC (ed): *Textbook of Neonatology.* Edinburgh, Churchill Livingstone, 1986; pp. 394–406.
12. Poland RL, Ostrea EM. *Neonatal Hyperbilirubinaemia.* In Klaus MH, Fanaroff AA (eds): *Care of the High Risk Neonate.* 2nd ed. Philadelphia, Saunders, 1979; pp. 253–60.
13. Roberton NRC. *A Manual of Neonatal Intensive Care.* 2nd ed. London, Edward Arnold, 1986; pp. 206–27.

26

Neonatal Screening for G6PD Deficiency and Congenital Hypothyroidism

Betty WY YOUNG
Vincent SM CHAN

1. Sample collection

1.1 Two specimens of cord blood are collected during deliveries
- ❏ 2.5 ml blood in plain specimen bottle for TSH assay by the immunoradiometric assay (IRMA).
- ❏ 2.5 ml blood in EDTA specimen bottle for quantitative assay of G6PD activity.

1.2 Specimens are sent to the Neonatal Screening Laboratory.

2. Laboratory investigation

2.1 Reference normal ranges
- ❏ Thyroid screening:

	TSH* (mIU/l)	T4 (nmol/l)	Free T4 (pmol/l)	T3 (nmol/l)
Cord blood	≤ 17	78–188	13.9–23.8	
1–3 weeks	≤ 17	110–254	20.6–38.2	1.5–3.5
3–4 months	≤ 17	95–200	20.6–38.2	
Adults	≤ 5	77–157	11.6–29.7	1.1–2.7

* The present cut off point for cord blood TSH screening is set at 17 mIU/l (2 SD from the mean).

❏ G6PD screening:
Glucose-6-phosphate dehydrogenase activity (U/gm Hb):

	Normal	Borderline normal	Deficient
Cord blood	4.3–9.0	1.7–4.2	≤ 1.6
Adults	2.7–6.0	1.1–2.6	≤ 1.0

3. Actions after screening

3.1 Results will be forwarded to the concerned obstetric units or maternity homes.

3.2 Follow-up of infants with abnormal results
❏ G6PD deficiency cases: Counselling given to parents or guardians.
❏ Screenees with elevated TSH will be subjected to the following evaluation:
- Clinical evaluation:

Birth weight > 4 kg	Cyanosis
Large anterior fontanelle	Goitre
Respiratory difficulty	Oedema
Abdominal distension	Hypothermia
Umbilical hernia	Constipation
Large tongue	Inactiveness
Sluggish jerks	Vomiting
Mottled skin	Poor feeding
Dry skin	Hoarse cry
Prolonged jaundice	

Maternal history of thyroid disease or medications
- Serum TSH & T4 rechecked after 5 days of postnatal life (after the postnatal TSH surge):
  – Normal results reassurance
  – TSH > 50 mIU/l further confirmatory tests
  – TSH 17–50 mIU/l or T4 < 110 nmol/l reassess at intervals
- Confirmation of congenital hypothyroidism (CHT):
  – Blood for TSH, T4, free T4, T3, TBG, antithyroglobulin and antimicrosomal antibodies
  – Bone age (knees +/– ankles for neonates)
  – Thyroid scintiscan
  – TRH stimulation test in selected cases with borderline TSH elevation

3.3 Treatment of confirmed congenital hypothyroidism with synthetic L-thyroxine

❑ Initial dose:
  10 mcg/kg/day (for first 3 months of age)
  (10 mcg/100 calories of estimated energy expenditure)
  (Alternative regime: 25 mcg T4 daily + 5 mcg T3 tid for 2 weeks
  and then stop T3)

❑ Maintenance doses:
  (100 mcg/square metre/day or 6 mcg/100 calories estimated energy expenditure)

0–6 months	25–50 mcg/day
6–12 months	50–75 mcg/day
1–5 years	75–100 mcg/day
6–12 years	100–150 mcg/day
> 12 years	100–200 mcg/day

❑ Check TSH and T4 levels one month after change of dosage.

3.4 Follow-up of infants with confirmed congenital hypothyroidism

❑ Patients are referred back to the paediatric units of their hospitals
  of birth or the hospitals of their residential regions.

❑ Suggested intervals of follow-up and blood testing:
  • One month after starting treatment
  • Every 3 months in the first year
  • Then every 6 months until school age
  • Then yearly till growth and development is complete

❑ Monitoring of progress:
  • Growth parameters
  • Psychomotor and intellectual development
  • Signs and symptoms of over- or under-treatment
  • Thyroid function tests: Maintain serum T4 in the upper half of
    the normal reference range
    (*Note*: Mild elevation of TSH may persist for some time.)

❑ Assessment for transient/permanent CHT:
  • Carried out at 3 years of age before committing patient to
    lifelong thyroxine replacement.
  • Patients with athyreosis or ectopic thyroids are excluded be-
    cause CHT is most likely to be permanent.
  • L-thyroxine (T4) is stopped and replaced with an equivalent
    dose of T3 (20 mcg T3 = 100 mcg T4) for 2 weeks followed
    by withdrawal of T3 for another 2 weeks.
  • Recheck thyroid function tests thereafter. Elevated TSH and
    decreased T4 confirms permanent CHT.
  • Thyroid scintiscan can be done at the same time in selected
    cases.

27
Techniques and Procedures

Paul KL LAM

1. Blood sampling

1.1 Capillary samples — heel stab
- ❏ Ensure adequate perfusion of heel by warming.
- ❏ Use only the medial and lateral surfaces of the plantar aspect of the heel (striped areas shown).

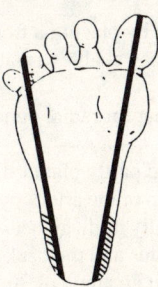

- ❏ Insert the stilette to a depth of 2.5 mm after sterilizing the puncture site with alcohol swab.
- ❏ Let blood ooze out, do not squeeze.
- ❏ Secure haemostasis after completion of sampling.

1.2 Venous blood
- ❏ Broken needle technique
  Break off the hub of an FG21 or FG23 needle and insert the needle into a peripheral vein in the hand or the foot.
- ❏ Venepuncture
  - It is safest to use the peripheral veins over limbs.
  - External jugular vein: The infant is held on his side with his head slightly lower than his trunk and with his neck extended

on the side to be entered. Use a butterfly with FG21 needle. Sit infant up and apply firm pressure after the procedure.

- Femoral vein (avoid as far as possible): If forced to use this site, hips must be fully abducted and adequately restrained. Puncture site must be thoroughly cleaned. Apply adequate pressure after the puncture to prevent haematoma formation.

1.3 Arterial puncture

7 sites can be used:

radial
ulnar
posterior tibial
dorsalis pedis
brachial ⎫
temporal ⎬ avoid as far as possible
femoral ⎭

❑ Radial:
- The wrist is held slightly extended.
- Locate the position of the artery by palpation.
- Insert the needle into the artery at an angle of 25–30° to the skin.
- Use FG23 or FG25 needle attached to a 1–2 ml heparinized syringe.
- Transillumination by placing a fibre-optic light source behind the baby's wrist may enable visualization of the artery.

❑ Ulnar:
Same technique as that for radial puncture.

❑ Posterior tibial:
- Baby's foot is held partly plantar flexed.
- Locate the position of the artery by palpation or transillumination which is usually midway between the posterior margin of the medial malleolus and the back of the achilles tendon.
- Insert the needle at an angle of 30–40° to the skin.

❑ Dorsalis pedis:
- Baby's foot is held partially plantar flexed.
- Locate position of artery by palpation or transillumination.
- Insert the needle at the midpoint of the foot where the artery usually lies between the first and second metatarsals.

❑ Brachial:
- Baby's arm is fully extended at the elbow and slightly externally rotated at the shoulder.
- Locate position of artery by palpation or transillumination.
- The artery is entered proximal to the antecubital fossa.

❑ Temporal:
- The artery runs along the lateral part of the skull and can be entered at any point along its course.

- Ascertain that the vessel is an artery by palpating for the arterial pulsation.
❑ Femoral:
  - This is the most undesirable site for arterial puncture.
  - Locate the position of the artery by palpation.
  - Enter the artery by a vertical approach about 0.5–1 cm distal to the inguinal ligament.

2. Percutaneous arterial catheterization

An FG23/24 indwelling catheter can be inserted percutaneously into the radial, ulnar, posterior tibial or dorsalis pedis arteries.

2.1 Before catheterizing the radial artery, confirm by Alan's test that the hand is adequately perfused by the ulnar artery alone (and vice versa for ulnar artery catheterization).

2.2 Never catheterize the brachial, temporal or femoral arteries.

2.3 The arterial catheter should be perfused with heparinized saline (2–5 units/ml) at a rate of 0.5 ml/hour. It can be left in situ for as long as it stays patent, but should not be used for anything other than blood sampling and arterial pressure monitoring.

3. Umbilical catheterization

Both the umbilical vein and artery can be catheterized for up to 7–10 days after delivery.

3.1 Insertion of an umbilical arterial catheter must be performed under full aseptic condition i.e. mask, gown and gloves.

3.2 After cleaning with antiseptic solution, a ligature should always be tied loosely around the base of the cord before the latter is cut.

3.3 The cord is cut about 5 mm distal to the skin-Wharton's jelly junction.

3.4 One of the arteries is gently dilated with a pair of fine forceps to a depth of 5–10 mm.

3.5 Hold the edge of the artery and surrounding Wharton's jelly with a pair of fine-toothed forceps, and introduce an FG4.5–5.0 PVC catheter full of heparinized saline (2 units/ml). The cord stump should be pulled up towards the infant's head to straighten out the artery as it turns caudally in the anterior abdominal wall just below the umbilicus.

3.6 Consult Table 1 for appropriate insertion distance.
2 locations to place the catheter tip:
❑ Above the diaphragm and below the ductus arteriosus, i.e. opposite vertebral bodies T_7 to T_{10}.

❏ Above the bifurcation of abdominal aorta and below the inferior mesenteric artery take off (L_4 to L_5).

3.7 Suture umbilical stump around the catheter in a purse-string fashion and secure the catheter with a plastic tape "bridge."

3.8 Sometimes an umbilical venous catheter is inserted into the right atrium to monitor CVP. Consult Table 1 for correct insertion distance.

Table 1. Insertion Distance for Umbilical Catheters (cm)

Shoulder (lateral end of clavicle) to umbilicus	Aortic catheter to diaphragm	Aortic catheter to aortic bifurcation	Venous catheter to right atrium
9	11	5	6
10	12	5	6–7
11	13	6	7
12	14	7	8
13	15	8	8–9
14	16	9	9
15	17	10	10
16	18	10–11	11
17	20	11–12	11–12

3.9 Precautions with umbilical catheter

❏ Catheter must be continuously flushed with heparinized saline or glucose/electrolyte solution.

❏ It can be kept for more than 2 weeks if necessary and should be removed when the infant no longer requires regular blood gas analyses.

❏ It has to be removed at once if blockade occurs or if the baby develops abdominal distension, overt N.E.C., haematuria or signs indicating circulatory impairment of the lower limbs (pallor, skin mottling or blackened areas over the buttocks, thighs or legs, and discoloured toes).

❏ Failure to pass the catheter is usually due to creation of a "false tract" by breaking through the intima of the vessel with or without penetrating into the surrounding tissues.

❏ Obtain abdominal and chest radiograph to confirm the proper position of the catheter. Reposition the catheter if necessary.

4. Lumbar puncture

4.1 Sterile preparation of the lower back is mandatory.

4.2 The procedure can be carried out with infant lying down or sitting up. Flex the lumbar spine as much as possible. Beware of compromising the airway on doubling up the infant.

4.3 Use the L3/4 or L4/5 space and a no. 22 gauge needle. Advance the needle for 5–7 mm (premature infants) to 1 cm (term infants) to reach the subarachnoid space. Often the operator cannot feel that subarachnoid space has been entered.

4.4 It may be advisable to delay the lumbar puncture in a critically ill newborn with unstable vital signs even if meningitis is suspected: treat the infant as such first and perform the lumbar puncture later when the infant is more stable.

5. Ventricular puncture

5.1 Always assess ventricular size ultrasonically or by CT before attempting a ventricular puncture. Attempts to puncture a non-dilated ventricle are often unsuccessful and hazardous.

5.2 The ventricles are entered by passing a fine LP needle through the lateral margin of the anterior fontanelle and aiming forward and slightly inwards towards the inner canthus of the opposite eye. (For babies with very wide fontanelle as in cases of hydrocephalus, insert the needle at the level of the eye.)

6. Subdural taps

6.1 Subdural space is entered at the lateral angle of the anterior fontanelle.

6.2 As soon as the "give" sensation is felt, which indicates the dura is penetrated, the stilette should be withdrawn.

References

1. Roberton NRC. *A Manual of Neonatal Intensive Care*. 2nd ed. Edward Arnold Ltd, 1986; pp. 295–306.

2. Hodson WA, Truog. *Critical Care of the Newborn*. 2nd ed. WB Saunders, 1989; pp. 184–85.

28

Drug

FOK Tai-fai

Drug	Route	Dosage	Notes
Acetazolamide (Diamox)	PO	7.5 mg/kg/dose q8h May increase dosage daily by 25 mg/kg/day up to a max of 100 mg/kg/day	For medical treatment of hydrocephalus. Side effect: metabolic (hyperchloraemic) acidosis — may require $NaCO_3$ treatment. May be given in combination with frusemide 1–3 mg/kg/day in 3 divided doses.
Acetylcysteine	ETT	0.2–0.5 ml before ETT suctioning	
Acyclovir (Zovirax)	IV infusion over 1H	10 mg/kg/dose q8h	For herpes simplex and varicellazoster virus infection.
Adrenaline (1/10000)	IV, ETT, IC	0.1 ml/kg/dose	For CPR.
Albumin 25% 5% (dilute with NS or D5)	IV	25%: 2–4 ml/kg 5%: 10–20 ml/kg	25%: For Hypoalbuminaemia or hyperbilirubinaemia. 5%: For volume replacement.
Amikacin (Amikin)	IV, IM	7.5 mg/kg/dose 1st wk: q12h > 1 wk: q8h Preterm: may need spacing out (q24h or even longer)	Blood levels: Peak: 15–20 µg/ml. Trough: 5–8 µg/ml.

Aminophylline	IV	Loading: 6.6 mg/kg/dose Continuous infusion: 4.4 mg/kg/24h	For neonatal apnoea. Blood level: 40–80 µmol/L ($\times$ 0.18 = mg/ml).
Amoxycillin (Amoxil)	IV, IM, PO	50 mg/kg/dose 1st wk: q12h 2–4 wk: q6h > 4 wk: q3h	
Amphotericin B	IV	Starting: 0.25 mg/kg/dose Maximum: 1 mg/kg/dose Infused over 6H as a single daily dose	Test dose: 0.1 mg/kg (up to a total of 1 mg) over 3–4h. Note pulse, temperature, respiration, BP. Then start with 0.25 mg/kg over 6h (may be given the same day as test dose). Increase by daily increments of 0.25 mg/kg over a 4-day period to reach a total dose of 1 mg/kg. Protect infusion bottle and tubings from light.
Ampicillin	IV, IM	50 mg/kg/dose < 7 days: q12h > 7 days: q8h Meningitis: 100 mg/kg/dose q6h	
ATP	IV	0.05 mg/kg/dose bolus q2 minimum to a maximum of 0.25 mg/kg	For supraventricular tachycardia.
Atropine	IV, IM, SC	0.01 mg/kg/dose	For CPR and premedication. Side effects: tachycardia, arrythmia, flushed dry skin, dilated pupils, tachypnoea, apnoea.
Augmentin (Amoxycillin + clavulanic acid)	IV, PO	30 mg/kg/dose < 4 wks: q12h > 4 wks: q6–8h	
Aztreonam (Azactam)	IV	20–40 mg/kg/dose q6h	

Drug	Route	Dosage	Notes
Calcium Chloride 10% (0.7 mmol Ca/ml) (1 ml = 27 mg Ca)	IV	CPR: 0.3 ml/kg/dose	Cardiac monitoring required during infusion.
Calcium Gluconate (10%) (0.22 mmol Ca/ml) (1 ml = 9 mg Ca)	IV	CPR: 0.1 ml/kg/dose Hypocalcaemic tetany: 2 ml/kg/dose Exchange transfusion: 1 ml/100 ml blood exchanged Replacement therapy: 9 ml/kg/day continuous infusion Daily maintenance: 4.5 ml/kg/day (1 mmol/kg/d) continuous infusion	Cardiac monitoring required during bolus infusion. Do not fash flush IV line during continuous infusion. Extravasation may result in very extensive and severe damage of tissues.
Carbenicillin	IV, IM	100 mg/kg/dose < 7 days: q8h (< 2 kg: 75 mg/kg q8h) > 7 days: q6h	
Carbimazole	PO	0.25 mg/kg/dose q8h × 2 wks then 0.1 mg/kg/dose q12h	For neonatal thyrotoxicosis.
Cefaclor (Ceclor)	PO	10–15 mg/kg/dose q8h	B.D. dose sufficient for UTI.
Cefadroxil Monohydrate (Duracef)	PO	25–100 mg/kg/day, BD or TDS	For respiratory tract infection, and UTI (E. coli, P. mirabilis, and Klebsiella species).
Cefamandole (Mandol)	IV, IM	25–40 mg/kg/dose q6–8h	
Cefoperazone (Cefobid)	IV, IM	25–50 mg/kg/dose q6–12h	Poorer stability to beta-lactamase and narrower spectrum than other 3rd generation cephalosporins. Excreted in bile — no need for dosage adjustment in renal impairment.

Drug	Route	Dose	Notes
Cefotaxime (Claforan)	IV, IM	25–50 mg/kg/dose < 7 days: q8h 2–4 wks: q6h > 4 wks: q4–6h	Wide spectrum *vs* enterobacter and H. influenza. Poor against staphylococci, B. fragilis and Ps. aeruginosa. Reduce dosage in renal failure. False +ve for glucose in urine.
Cefoxitin (Mefoxin)	IV	20–40 mg/kg/dose q6h	
Ceftazidime (Fortum)	IV, IM	25–50 mg/kg/dose q12h Severe infection, e.g. meningitis: 50 mg/kg q8h	Broad spectrum. Effective against Ps. aeruginosa. Reduce dosage in renal failure.
Ceftriaxone (Rocephin)	IV, IM	50 mg/kg/dose q24h Meningitis: loading dose 75 mg/kg, then 50 mg/kg/dose q12h	Excreted in bile. May result in biliary stone formation.
Cefuroxime (Zinacef)	IV	20–50 mg/kg/dose q12h Severe infections: 50 mg/kg/dose q6h	Effective *vs* many enterobacteria and streptococcus, staphylococcus and H. influenza. Reduce dosage in renal impairment.
Cephalexin (Keflex)	PO	10–25 mg/kg/dose q6h	
Cephalothin (Keflin)	IV, IM	20 mg/kg/dose q6h	
Cepharadine (Velosef)	IV, IM, PO	10–25 mg/kg/dose q6h	
Chloral hydrate	PO	Hypnosis: 50 mg/kg/dose Sedation: 25 mg/kg/dose Maintenance: 6–12 mg/kg/6–8h	May cause gastric irritation: give with milk or water feed.
Chloramphenicol	IV, IM, PO	Loading dose: 40 mg/kg Then 25 mg/kg/dose Preterm (up to 4 wk): q24h Term: < 7 days: q24h > 7 days: q12h	Monitor blood levels: Peak: 20–30 µg/ml. Trough: < 15 µg/ml.

Drug	Route	Dosage	Notes
Chlorothiazide	PO	10–20 mg/kg/dose q12–24h	Monitor serum electrolytes. May cause Hypothemia tachycardia, cholopytatic jaundice and skin rash.
Chlorpromazine	PO, IM	0.5–1 mg/kg/dose q6h	For sedation (narotic withdrawal).
Cimetidine (Tagamet)	IV, PO	5–10 mg/kg/dose q6h	For gastric bleeding. Antagonizes action of tolazoline in PFC, increases serum levels of phenytoin and theophylline.
Clonazepam (Rivotril)	IV, PO	Status: 0.25 mg/kg/dose IV (may be repeated) Maintenance: 0.01–0.06 mg/kg/dose q8h orally	For infantile spasm or other myoclonic seizures.
Cloxacillin	IV, PO	25–50 mg/kg/dose < 7 days: q12h 2–4 wks: q8h > 4 wks: q4–6h	
Cotrimoxazole (trimethoprim: sulphamethoxazole = 1:5) (Septrin, Bactrim)	IV, PO	Trimethoprim: 2.5 mg/kg/dose q12h Pneumocystis: 5–10 mg/kg/dose q6h	
Cryoprecipitate (each ml contains: 25 mg fibrinogen, 5–10 u factor 8)	IV	1 unit/kg factor 8 increases activity by 2%, half life 12H	

Drug	Route	Dose	Notes
Dexamethasone	IV, IM	Airway oedema: 0.25 mg/kg/dose q6h Cerebral oedema: Initial = 0.5–1 mg/kg (max.: 10 mg) then: 0.1–0.2 mg/kg/dose q6h × 5 days BPD: 0.5 mg/kg/dose q12h × 3 days then 1/2 dose × 3 days then 1/4 dose × 3 days then 1/4 dose qd × 3 days	Ensure that patient is free from sepsis when dexamethasone is given.
Dextrose (glucose)	IV	Hypoglycaemia: 1 ml D25 or D50/kg Hyperkalaemia: Insulin 0.1 u/kg + D50 2 ml/kg	
Diazepam (Valium)	IV	0.2 mg/kg/dose	For convulsion resistant to other anticonvulsants. Avoid in jaundiced neonates.
Digoxin	PO, IV	TDD: 0.04 mg/kg 1/2 stat 1/4 at 8h 1/4 at 16h Maintenance (8H after TDD): 0.01 mg/kg/day	Toxicity common — vomiting, diarrhoea, arrhythmia. Toxicity enhanced by potassium or calcium imbalance, diuretics, etc. May need to reduce dose in preterms. Serum level: 1.3–2.6 nmol/L (× 0.78 = ng/ml)
Dobutamine	IV	2.5–10 µg/kg/min	No direct effect on renal perfusion; little or no peripheral dilatation or tachycardia. May be given together with dopamine.

Drug	Route	Dosage	Notes
Dopamine	IV	2–20 µg/kg/min	At low dose: peripheral vasodilatation, minimal renal arterial vasodilatation. At dose > 10 µg/kg/min (or even lower doses in neonates): alpha adrenergic effect predominant (decreased renal perfusion). Stop infusion if tachycardia, arrhythmia, or oliguria develops. Incompatible with $NaHCO_3$.
Doxapram	IV infusion	Initial: 1 mg/kg/H Then 0.5–2.5 mg/kg/H	For neonatal apnoea. May cause severe hypertension.
Erythromycin	IV (infusion ~ 30 min or continuous) PO	10–12 mg/kg/dose q6–8h	IM injection painful. Hepatotoxic, cholestasis.
Ethambutol	PO	10–15 mg/kg/dose, daily	Side effects: May cause optic neuritis — need frequent ophthalmological examination.
Ethionamide	PO	7.5 mg/kg/dose BD	Side effects: GI upset, hepatotoxic (jaundice and raised GOT), neurotoxic (neuritis), rash.
Fentanyl	IV	1–2 µg/kg/dose Ventilated: 5–10 µg/kg/dose	
Ferrinsol (15 mg Fe/0.6 ml)	PO	0.3 ml/dose, QD-BD	
Flucloxacillin	IV, IM, PO	25 mg/kg/dose < 2000 g: q12h > 2000 g: < 15 days: q8h > 15 days: q6h	Better absorbed than cloxacillin when given enterally.

Drug	Route	Dose	Notes
Flucytosine (fluorocytosine)	IV (infusion ~ 30 min) PO	50 mg/kg/dose 6–8H	Hepatotoxic.
Folic acid	IV, IM, PO	Treatment: 0.25 mg/kg daily	
Fresh frozen plasma	IV	10–20 ml/kg	Contains all clotting factors.
Frusemide (Lasix)	IV, IM, PO	1–3 mg/kg/dose q12–24h	Total daily dose not to exceed 6 mg/kg. Monitor serum electrolytes in prolonged therapy.
Fusidic acid	IV (infusion ~ 30 min) PO	20 mg/kg/dose start then 10–15 mg/kg q8h	Effective vs MRSA. Synergistic with erythromycin & lincomycin. Bacterial resistance readily develops when used as a single drug. Blood level 40–400 μmol/ (× 0.52 = μg/ml)
Gentamycin	IV, IM	2.5 mg/kg/dose <7 days: q12h >7 days: q8h Preterm: may need spacing out to q24h	Monitor blood level: Peak: 5–10 μg/ml. Trough: < 2 μg/ml.
Glucagon	IV	0.2 μg/kg stat Then 0.005–0.01 μg/kg/H (1 μ = 1 mg)	For treatment of acute refractory hypoglycaemia not responding well to glucose therapy.
Heparin	IV	Vascular line (prevent clotting): 1 μ/ml Heparinizing fresh blood for transfusion/ exchange transfusion: 2 μ/ml	Protamine sulphate required after exchange transfusion with heparinized fresh blood.
Hydrochlorothiazide	PO	1–4 mg/kg/dose BD	
Imipenem (Tienam)	IV	15–25 mg/kg/dose q6h	Monitor serum electrolytes.
Immunoglobulin (Human)	IV (infusion ~ 24H)	500–1000 mg/kg q14 days	Various dosage regimes have been recommended.

Drug	Route	Dosage	Notes
Immunoglobulin Hepatitis B (Hyperhep)	IM	100 u (0.5 ml) given to babies of HBV carrier mothers within 24H of birth	
Immunoglobulin, Tetanus	IM	250–500 units (prophylaxis)	
Indomethacin	IV, PO	0.2 mg/kg/dose q12H for 3 doses OR 0.1 mg/kg/dose q24H for 6 doses	For closure of PDA. Contraindications: bleeding tendency, NEC, significant NNJ, or severe renal impairment.
Insulin	IV	Hyperglycaemia: 0.05–0.2 μ/kg/dose prn or 0.1 μ/kg/H in 5% albumin or plasma; Hyperkalaemia: insulin 0.1 μ/kg + D50 2 ml/kg continuous infusion over 2H	Monitor "dextrostix" and blood glucose closely
Ipratropium (Atrovent)	Nebulized (250 μg/ml)		Anti-cholinergic (atropine-like). Used alone or in combination with other bronchodilators on babies with BPD.
Isoniazid	PO	10–15 mg/kg/dose daily TB meningitis: 20 mg/kg/day	Side effects: peripheral neuropathy, convulsion. Add pyridoxine (10 mg of pyridoxine for every 100 mg isoniazid).
Isoprenaline	IV infusion	Initial dose: 20 μg/kg/H Maintenance: Double infusion rate every 2 min until response	Indication: hypotension due to impaired cardiac output. Incompatible with any other drug in solution.
Ketoconazole	PO	5 mg/kg/dose q12–24h	Hepatotoxic (idiosyncracy)
Lignocaine	IV	1 mg/kg/dose — may be repeated every 10 min for up to 6 doses Continuous infusion: 0.5–1.5 mg/kg/H	For ventricular ectopics. Ventricular tachydysrhythmias, digitalis induced ventricular arrythmia. Side effects: dyspnoea, respiratory arrest. Toxic effects: drowsiness, convulsion.

Magnesium sulphate (50%) (magnesium 2 mmol/ml)	IM	0.5 ml/kg/dose	For hypomagnesaemia.
Magnesium trisilicate	PO	1–2 ml/dose	For gastric bleeding (stress ulcer).
Mannitol	IV	0.25–0.5 g/kg/dose (1–2 ml/kg 25% solution) q2h prn	Stop giving if serum osmolality > 330 mosm/L.
Methylene blue	IV	1–2 mg/kg/dose	For methaemoglobinaemia.
Metronidazole (Flagyl)	IV	15 mg/kg stat, then 7.5 mg/kg/dose < 1 wk: q12h 2–4 wk: q8h > 4 wk: q6h	For anaerobic infections.
Miconazole	IV (infusion ~ 1H)	7.5–15 mg/kg/dose q8h	Side effects: transient decrease in Hct and platelet. Vomiting, diarrhoea.
Morphine	IV, IM	IM: 0.2 mg/kg/dose q4–6h prn IV: 0.1–0.2 mg/kg/dose q4–6h prn IV infusion: 0.5 mg/kg in 50 ml heparin-D5 at 1–4 ml/H (10–40 µg/kg/H)	
Moxalactam (Moxam)	IV, IM	50 mg/kg/dose < 7 days: q12h > 7 days: q8h	Patient should be given Vit. K to prevent coagulopathy.
Naloxone (Narcan)	IV, ETT, IM, SC	0.01 mg/kg/dose Repeat prn 0.01 mg/kg	For treating respiratory depression due to narcotic analgesics.
Neomycin	PO	12.5–25 mg/kg/dose q6h	

Drug	Route	Dosage	Notes
Netilmicin (Netromycin)	IV, IM	2.5 mg/kg/dose < 7 days: q12h > 7 days: q8h	Monitor blood level: Peak: 10–12 µg/ml. Trough: 2–4 µg/ml.
Nitrofurantoin	PO	1.5 mg/kg/dose q6h UTI prophylaxis: 2.5 mg/kg at night	
Nystatin	PO	100,000 units (1 ml)/dose q6–8h	
Packed cells	IV	10 ml/kg raises Hb by 3g% 1 ml/kg raises PCV by 1%	
Pancuronium (Pavulon)	IV	0.1 mg/kg/dose Repeat prn	Do not mix with other drugs.
Paraldehyde	IM, IV, Rectal	IM: 0.2 ml/kg stat, then 0.1 ml/kg/dose q4–6h IV: 0.2 ml/kg over 15 min, then 0.02 ml/kg/H Rectal: 0.3 ml/kg/dose diluted with equal volume of liquid paraffin	Use only glass or special syringe; change syringe and tubing 8 hourly if given rectally.
Penicillin G (benzyl)	IV	50,000–100,000 units/kg/dose (30–60 mg) < 7 days: q12h > 7 days: q6h GBS meningitis: 100,000 µg/kg q6h	
Pethidine	IV, IM	IM: 1–1.5 mg/kg/dose q4–6h prn IV: 0.5–1 mg/kg/dose q4–6h prn IV infusion: 5 mg/kg in 50 ml heparin-D5 at 1–3 min/H (100–300 µg/kg/H)	Contraindicated in hepatic failure.

Phenobarbitone (Luminol)	IV, IM, PO	Loading dose for status convulsion: 20–30 mg/kg/dose IV Maintenance: 1.5–4 mg/kg/dose q12h	Therapeutic blood level: 65–130 μmol/L. Overdose may cause hypoventilation, hypotension, hypothermia and renal failure.
Phenytoin (Dilantin)	IV, PO	Loading dose for status convulsion: 15–20 mg/kg/dose over 30 min IV Maintenance: 3 mg/kg/dose q12h	Therapeutic blood level: 40–80 μmol/L.
Piperacillin (Pipracil)	IV over 3–5 min	50 mg/kg/dose < 7 days: q6–8h > 7 days: q4–6h	For serious infections e.g. E. coli, pseudomonas Aeruginosa. Should not be mixed with aminoglycoside in the same syringe. Adjust dosage in renal impairment.
Platelet concentrate	IV	10 ml/kg/dose (1 unit/10 kg will increase platelet by 40×10^9/L)	
Potassium	IV.	Maximum 0.5 mmol/kg/H IV Daily requirement: 2–4 mmol/kg 1 g KCL = 13.3 mmol potassium 15% KCL: 1 ml = 2 mmol potassium	
Propylthiouracil	PO	3 mg/kg/dose q8h	For neonatal thyrotoxicosis.
Prostaglandin E1 or E2	IV, PO	IV: 0.05–0.1 μg/kg/min continuous infusion; half dose once stabilized PO: 10–70 μg/kg/dose q1–4h	For ductus dependent heart diseases. Also decreases pulmonary vascular resistance. Side effects: apnoea (monitor patient closely), fever, vasodilatation, diarrhoea).
Protamine	IV	1 mg/100 units heparin	May cause systemic hypotension.
Pyridoxine	IV, PO	Test dose: 50 mg IV Maintenance: 2–5 mg/kg/day iv/po	For pyridoxine-dependent convulsions.

Drug	Route	Dosage	Notes
Ranitidine (Zantac)	IV (slow), PO	IV: 1 mg/kg/dose q6–8h PO: 3 mg/kg/dose q8–12h	For gastric bleeding (stress ulcer). Dilute drug with NS 1:4 and inject slowly over 5–10 min.
Ribavirin (6 g/vial)	Nebuilized PO	Nebulized: 1 vial (at conc of 20 mg/ml) for 12–18H daily for at least 3 days, max 7 days, by SPAG-2 aerosal generator PO: 3 mg/kg/dose q6–8h, for 5–14 days	Teratogenic effect unknown.
Rifampicin	PO	10–20 mg/kg single daily dose	Hepatotoxic, GI upset. May colour urine, tears, sweat, sputum and saliva red.
Salbutamol (Ventolin)	Nebulized	Respirator solution (0.5%): 0.05 ml/kg/dose diluted to 2 ml 3–6H	
Sodium (1 ml 30% NaCl = 5 mmol)	IV, PO	Requirement: 2–6 mmol/kg/d Depletion: weight × 0.3 × (140-serum Na)	
Spironolactone (Aldactone)	PO	1–1.5 mg/kg/dose q12h	For heart failure and BPD. Usually combined with frusemide or thiazide.
Sucralfate (1 g tab)	PO	1/4 tab per dose q6h	For stress ulcer.
Theophylline (Nuelin)	PO	5 mg/kg stat, then 2 mg/kg/dose q12h	For apnoea of prematurity. Therapeutic serum level: 40–80 µmol/L.
Thiopentone	IV	Loading 3–5 mg/kg slow infusion Then 1–4 mg/kg/H	Therapeutic serum level: 150–200 µmol/L. Beware of hypotension.

Thyroxine (T4)	PO	For congenital hypothyroidism: Start with 10 µg/kg single daily dose, gradually increasing to 50–75 µg/day depending on blood TSH + T4 or sTSH levels	
Ticarcillin (Ticarpen)	IV infusion over 30 min	50 mg/kg/dose < 7 days: q6–8h > 7 days: q4–6h	For severe infections especially Ps. aeruginosa, proteus, and anaerobes. Reduce dose in renal failure.
Tolazoline	IV	1–2 mg/kg slowly stat, then 1–2 mg/kg/H (max up to 10–15 mg/kg/H)	For PFC Causes severe hypotension — B.P. monitoring mandatory, and volume expander should be ready. Discontinue if no response after 60 min of infusion. Watch out for GI haemorrhage.
Tubocurarine	IV	0.6 mg/kg/dose prn	May cause systemic hypotension (peripheral vasodilatation). If patient hypotensive, pancuronium preferred.
Vancomycin (Vancocin)	IV (infusion ~ 30 min)	15 mg/kg/dose < 7 days: q12h > 7 days: q8h	Rapid infusion may cause "red man syndrome." Monitor blood levels: Peak: 20–40 mg/L (14–28 µmol/L). Trough: 5–10 mg/L (3–7 µmol/L).

Drug	Route	Dosage	Notes
Verapamil (Isoptin)	IV, PO	IV: 0.075–0.2 mg/kg/dose over 10 min. May repeat same dose after 2–5 min. Then 5 µg/kg/min. if necessary. PO: 4–10 mg/kg/day in 3–4 divided doses.	Bolus injection to be given under ECG monitor. For SVT, atrial flutter, AF, IHSS, angina. Contraindicated in severe LV failure, sick sinus syndrome, AV block. Potentiates AV block in patients receiving beta blockers or digitalis. Side effects: hypotension, bradycardia, asystole.
Vimax (Vit. A, D, C, B complex)	PO	0.3 ml BD	
Vitamin D: 1, 25-OH-D3 (Rocaltrol) 1-OH-D2 (dihydrotachysterol)	PO PO	Start 0.01 µg/kg/day 20 µg/kg/day	For hypoparathyroidism or renal ricket.
Vitamin E (Aquasol E)	PO	Preterms: 15 iu (0.3 ml) daily (1 iu = 1 mg)	No conclusive evidence on the beneficial effects of routine administration to preterm infants.
Vitamin K	PO, IM	PO: 1–2 mg IM: 0.5–1 mg	For prophylaxis in newborns.

Appendix 1

Growth Parameters

Growth Chart: Neonatal

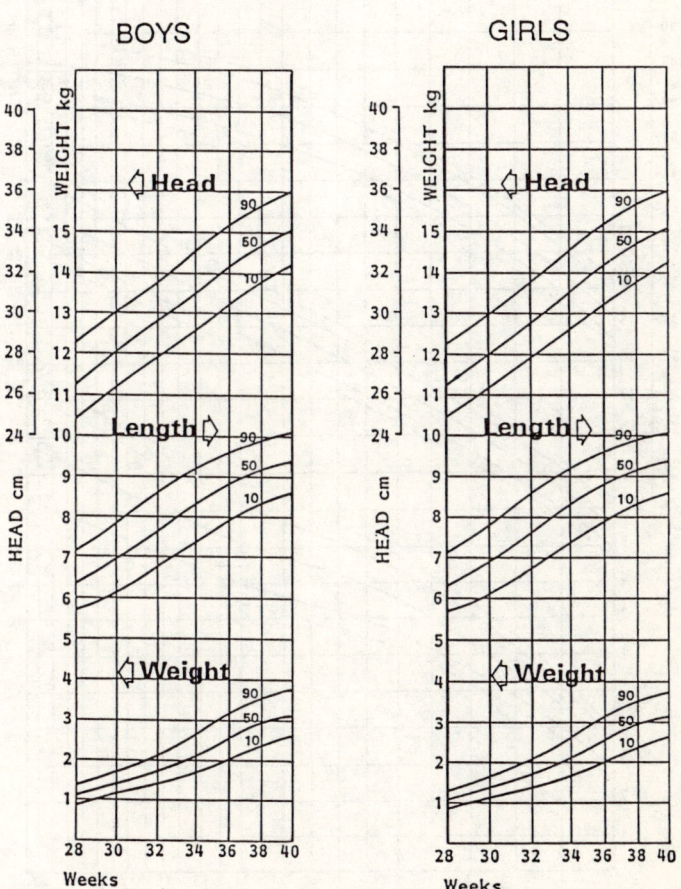

BOYS

GIRLS

Fok TF, Lam TK, Lee N, Chow CB, Au Yeung HCL, Leung NK and Davies DP. *A prospective study in the intrauterine growth of Hong Kong Chinese babies*. Biology Neonate 1987; 51:312–23.

Growth Chart: Postnatal to 2 Years

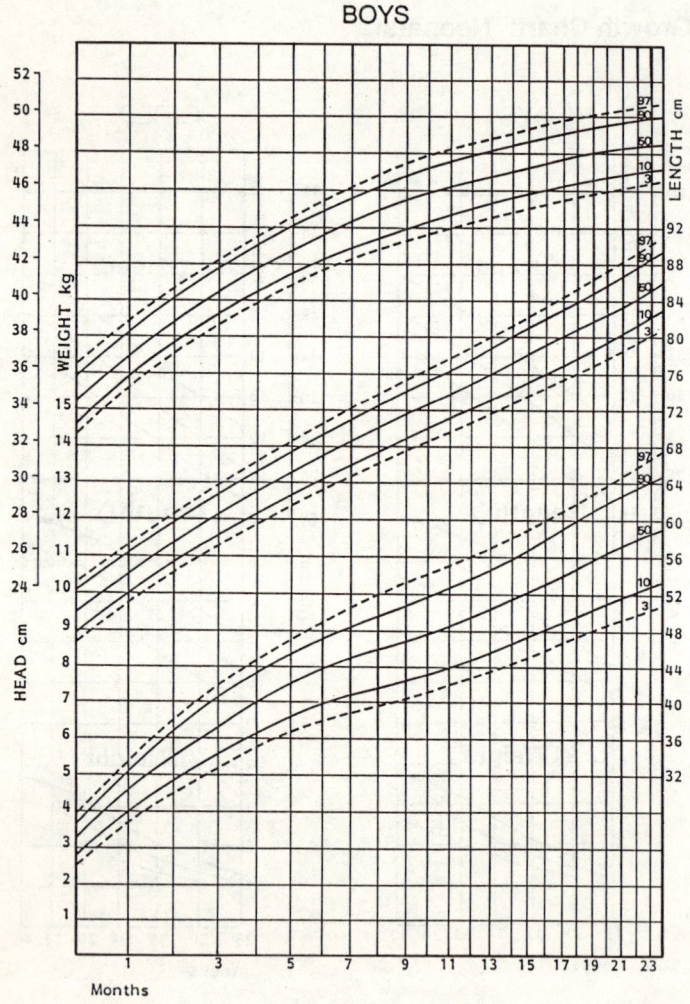

Leung SSF, Lui S, Lo L, Lam YM and Davies DP. *Growth standards for weight, length and head circumference: Hong Kong infant birth – 2 years.* Hong Kong Journal of Paediatrics 1988; 5:109–24.

Growth Chart: Postnatal to 2 Years

GIRLS

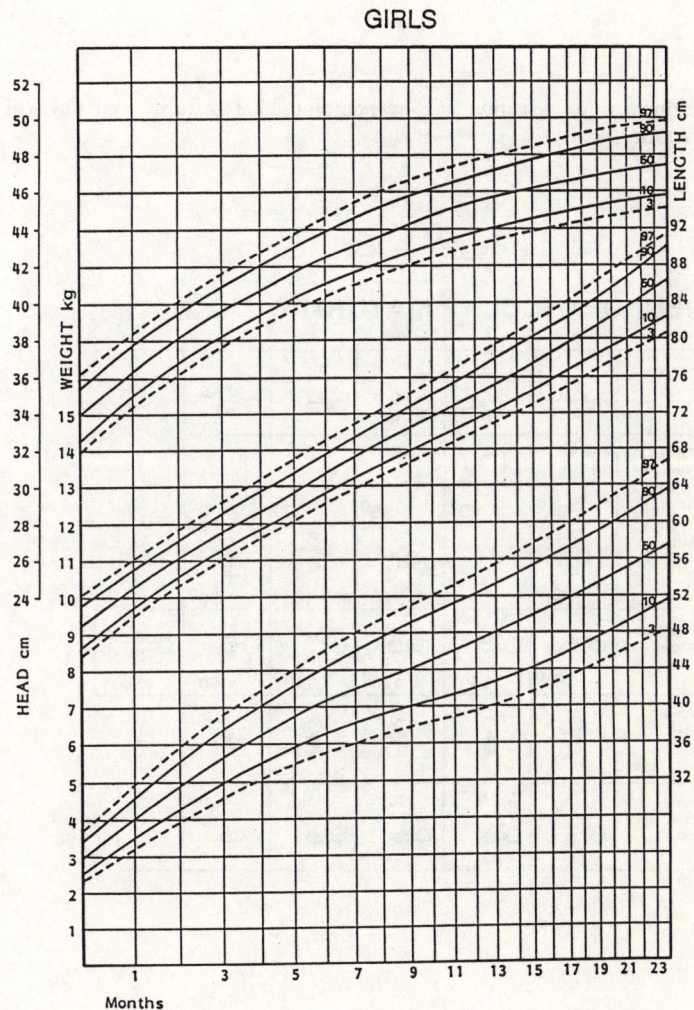

Months

Leung SSF, Lui S, Lo L, Lam YM and Davies DP. *Growth standards for weight, length and head circumference: Hong Kong infant birth – 2 years.* Hong Kong Journal of Paediatrics 1988; 5:109–24.

Appendix 2
Assessment of Newborn Maturity

Estimation of gestation age by neuromuscular maturity and physical maturity.

NEUROMUSCULAR MATURITY

	0	1	2	3	4	5
Posture						
Square Window (Wrist)	90°	60°	45°	30°	0°	
Arm Recoil	180°		100°-180°	90°-100°	< 90°	
Popliteal Angle	180°	160°	130°	110°	90°	< 90°
Scarf Sign						
Heel to Ear						

PHYSICAL MATURITY

	0	1	2	3	4	5
SKIN	gelatinous red, trans-parent	smooth pink, visible veins	superficial peeling &/or rash, few veins	cracking pale area, rare veins	parchment, deep cracking, no vessels	leathery, cracked, wrinkled
LANUGO	none	abundant	thinning	bald areas	mostly bald	
PLANTAR CREASES	no crease	faint red marks	anterior transverse crease only	creases ant. 2/3	creases cover entire sole	
BREAST	barely percept.	flat areola, no bud	stippled areola, 1–2 mm bud	raised areola, 3–4 mm bud	full areola, 5–10 mm bud	
EAR	pinna flat, stays folded	sl. curved pinna, soft with slow recoil	well-curv. pinna, soft but ready recoil	formed & firm with instant recoil	thick cartilage, ear stiff	
GENITALS **Male**	scrotum empty, no rugae		testes descend-ing, few rugae	testes down, good rugae	testes pendulous, deep rugae	
GENITALS **Female**	prominent clitoris & labia minora		majora & minora equally prominent	majora large, minora small	clitoris & minora completely covered	

Maturity rating score

Score	5	10	15	20	25	30	35	40	45	50
Weeks	26	28	30	32	34	36	38	40	42	44

Scoring system: Ballard JL, et al. *A simplified assessment of gestational age.* Pediatr Res. 1977; 11:374.
Figures adapted from: Sweet AY. *Classification of Low-birth-weight Infant.* In Klaus MH and Fanaroff AA (eds): *Care of the High-Risk Infants.* Philadelphia, WB Saunders, 1977; p. 47.

Appendix 3

Length for Umbilical Artery Catheterization

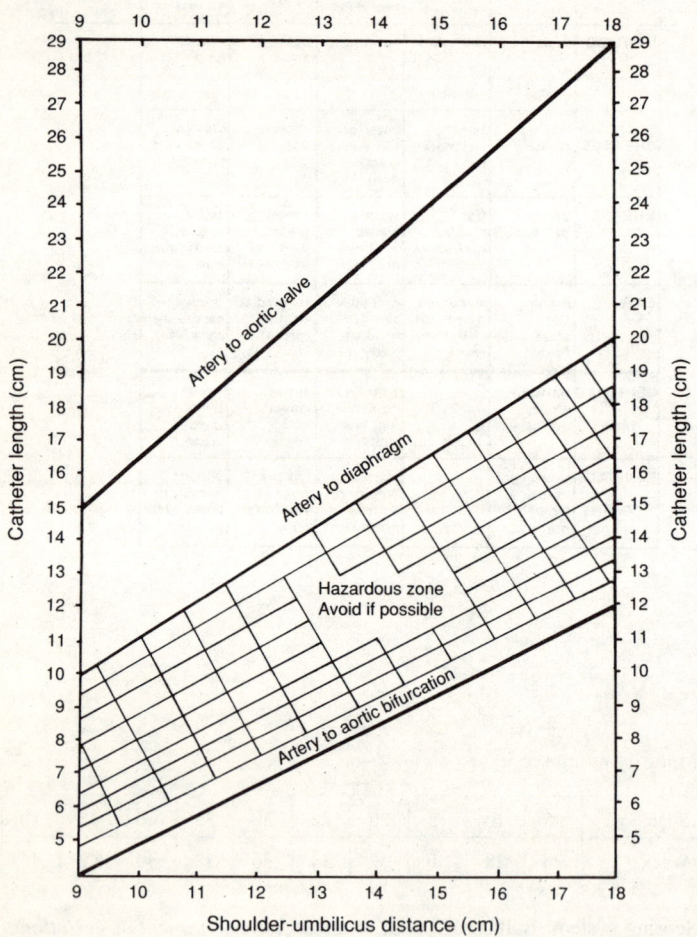

Distance from shoulder to umbilicus measured from above the lateral end of the clavicle to the umbilicus, as compared with the length of the umbilical artery catheter needed to reach the designated level.

Appendix 4

Length for Umbilical Vein Catheterization

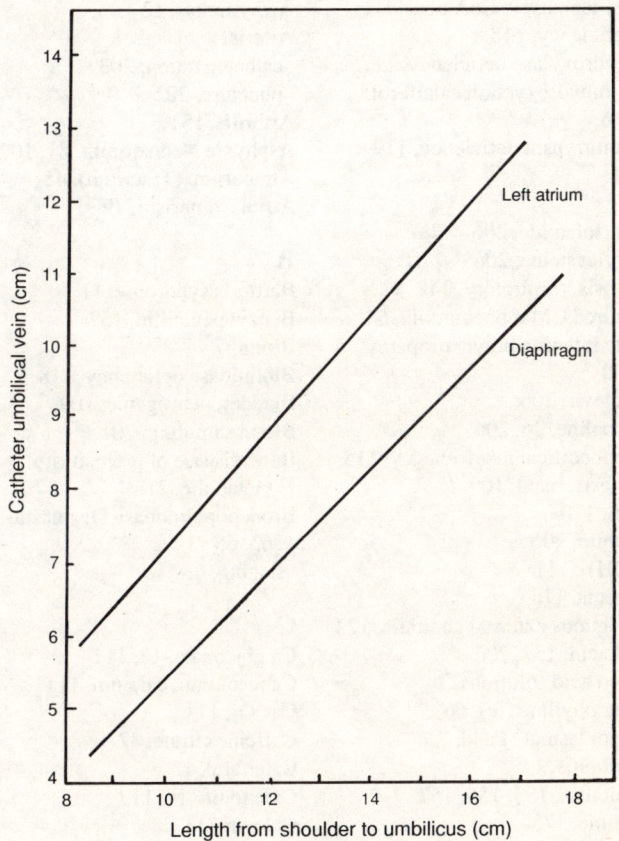

Catheter length for umbilical vein catheterization. The catheter tip should be placed between the diaphragm and the left atrium.

Index